INDUSTRIAL DRYING EQUIPMENT

CHEMICAL INDUSTRIES

A Series of Reference Books and Textbooks

Consulting Editor
HEINZ HEINEMANN
Heinz Heinemann, Inc.,
Berkeley, California

Additional Volumes in Preparation

INDUSTRIAL DRYING EQUIPMENT

EQUIPMENT

Selection and Application

C. M. van 't Land

Akzo Chemicals B.V.
Deventer, The Netherlands

Marcel Dekker, Inc.   New York • Basel • Hong Kong

Library of Congress Cataloging-in-Publication Data

Land, C. M. van 't
    Industrial drying equipment : selection and application / C. M.
van 't land
        p.   cm. -- (Chemical industries; v. 45)
    Includes bibliographical references and index.
    ISBN 0-8247-8316-6
    1. Drying apparatus. I. Title. II. Series.
TP363.L23   1991
660'.28426--dc20   91-16945
                                                                      CIP

This book is printed on acid-free paper.

Marcel Dekker, Inc.
270 Madison Avenue, New York, New York 10016

Current printing (last digit):
10 9 8 7 6 5 4 3 2 1

PRINTED IN THE UNITED STATES OF AMERICA

Preface

Drying is an important unit operation in the process industry. This book treats drying as a method for accomplishing a liquid/solid separation by other than mechanical means. Usually, heat is supplied, leading to evaporation of a liquid (usually water), and this leaves a solid behind. Drying accomplishes the transformation of a process stream and, as such, produces a salable product. As drying is an energy-intensive activity and dryers are expensive pieces of equipment, drying must be carried out as economically as possible.

Among those interested in drying are chemical engineers, energy specialists, and mechanical engineers. The objective of this book is to assist the process development engineer, the process engineer, and the plant engineer in their selection of drying equipment. The criteria to be observed are discussed, as are the means to estimate the financial consequences of a choice. Procedures for sizing equipment are also covered.

This book is primarily practical in nature, but it can also be used in conjunction with available theoretical books.

C. M. van 't Land

Acknowledgments

The writing of this book has been made possible by permission of Akzo Chemicals B.V., to whose management I am especially grateful. The invaluable experience gained while in their employ has been an important element in the design of this book.

Thanks are due to D. J. Buckland of Akzo Chemicals Ltd. for his help in converting my "Dutch English" into proper English as well as for suggesting a number of improvements to the contents. I am grateful as well to J. W. Postma who read the manuscript on behalf of Akzo Chemicals B.V. and, in so doing, made useful suggestions.

Particular appreciation is expressed for the assistance given by:

F. A. Aiken, SWENSON: rotary dryer
M. W. Cannon, APV Pasilac: centrifugal fluid-bed dryer
J. J. de Groot, Akzo Chemicals: process safety
B. A. S. Letham, Protor & Schwartz: conveyor dryer
K. Masters, Niro Atomizer: spray dryer
M. Matter, Buss: Convex dryer and vigorously agitated contact dryer
H. Motek, Salzgitter Maschinenbau: Rapid dryer
J. W. A. Renssen, Goudsche Machinefabriek: drum dryer
J. E. Starkowski, Combustion Engineering: flash dryer

The author thanks the following companies that most kindly provided data, drawings, and/or photographs:

Adolf Kühner AG, Birsfelden, Switzerland
APV Pasilac Limited, Carlisle, England
Babcock-BSH AG, Krefeld-Uerdingen, Federal Republic of Germany
Babcok-Duiker bv, Kwintsheul, The Netherlands

Bépex Corporation, Minneapolis, Minnesota
Brabender Messtechnik KG, Duisburg, Federal Republic of Germany
Buell Operation, Birmingham, England
Buss AG, Basel, Switzerland
Carrier Vibrating Equipment, Inc., Louisville, Kentucky
CSM Suiker b.v., Diemen, The Netherlands
Edwards High Vacuum International, Crawley, England
Euro–Vent Limited, Stoke-on-Trent, England
Fenwal Incorporated, Ashland, Massachusetts
FMC Corporation, Colmar, Pennsylvania
Gala Industries, Inc., South Charleston, West Virginia
General Eastern Instruments, Watertown, Massachusetts
Glatt GmbH, Binzen, Federal Republic of Germany
Goudsche Machinefabriek B.V., Gouda, The Netherlands
Graviner Limited, Colnbrook, England
Hosakawa-Nauta BV, Haarlem, The Netherlands
Industrial Heating Benelux, Breda, The Netherlands
Kay-Ray Inc., Wheeling, Illinois
Koch Engineering Company Inc., Wichita, Kansas
Komline-Sanderson Engineering Corporation, Peapack, New Jersey
Krauss Maffei Verfahrenstechnik GmbH, Munich, Federal Republic of Germany
Laboratorium Prof. Dr. Berthold, Wildbad im Schwarzwald, Federal Republic of
    Germany
Leybold AG, Cologne, Federal Republic of Germany
Mahlo GmbH & Co KG, Saal/Donau, Federal Republic of Germany
Microdry Corporation, San Ramon, California
Moisture Systems Corp., Hopkinton, Massachusetts
Niro Atomizer, Soeborg, Denmark
Pacer Systems, Inc., Billerica, Massachusetts
Patterson-Kelley Co., East Stroudsburg, Pennsylvania
Proctor & Schwartz, Inc., Horsham, Pennsylvania
Proctor & Schwartz Ltd, Glasgow, England
Raymond Division, Combustion Engineering, Inc., Chicago, Illinois
Rosenmund AG, Liestal, Switzerland
Salzgitter Maschinenbau GmbH, Münster, Federal Republic of Germany
Schugi Process Engineers, Lelystad, The Netherlands
Stordy Combustion Engineering Ltd, Wolverhampton, England
Strayfield International Limited, Reading, England
Sulzer Escher Wyss GmbH, Ravensburg, Federal Republic of Germany
SWENSON INC., Harvey, Illinois
Werner & Pfleiderer, Stuttgart, Federal Republic of Germany

I wish to thank the following publishers, who most kindly provided
permission to use material:

Butterworth Scientific Ltd, Guildford, England
Informations Chimie, Paris, France

McGraw-Hill Book Company, New York, New York
VDI-Verlag GmbH, Düsseldorf, Federal Republic of Germany

Once again, I am greatly indebted to my wife, Annechien, for her constant encouragement and invaluable help in typing and assembling the manuscript.

Contents

1

INTRODUCTION

Drying can be defined as a unit operation in which a liquid/solid separation is accomplished by the supply of heat, with separation resulting from the evaporation of liquid. Although in the majority of cases water is the liquid being removed, solvent evaporation is also encountered. The definition may be extended to include the dehydration of food, feed, and salts and the removal of hydroxyl groups from organic molecules.

This book is based on personal experience gained in the selection of drying equipment while employed by Akzo, a multinational company that manufactures fibers, bulk and fine chemicals, pharmaceuticals, and coatings.

During the early stages involving the selection of a new dryer or the replacement of one, a person should seek the cooperation of a reputable dryer manufacturer. Such close cooperation between the manufacturer and the potential user is essential, because one partner is knowledgeable about the equipment and the other person has expertise in the product. Since small-scale testing of drying equipment can be performed, this procedure can give very valuable insight into ultimate dryer selection. However, it is important that the partners have some insight into the other's field. Thus, the user can make value judgments on the equipment being recommended by the manufacturer. The size of the equipment must be checked, using various techniques (estimating methods, rule of thumb, rough-and-ready calculations, etc.). This book covers these techniques for each class of dryer, with some computer programs for screening purposes also included.

Various reasons exist for drying materials to a specific level or range:

1. It is often necessary to obtain a free-flowing material that can be packed, transported, or dosed.
2. Contractual limits exist for some products, e.g., salt, sand, and yarn.
3. Statutory limits are in force for some materials, e.g., tobacco and flour.
4. A moisture content within a specified range may have to be obtained for quality reasons. For many dried foods, too much moisture may adversely affect shelf life and nutritional value, whereas a moisture content too low may make the product less enjoyable and the over-drying may cause the loss of valuable nutrients.
5. The efficiency of subsequent process steps sometimes requires the moisture content to be between specified limits, as in, e.g., the milling of wheat, or the pressing of pharmaceutical tablets.
6. The onset of mildew and bacterial growth in textiles, such as woolen cloth, can be prevented by drying the cloth to a specific moisture content.
7. For the manufacture of ready-to-use pottery articles.

Heat required for drying can be supplied by the fundamentally different mechanisms of convection, conduction, and radiation.

1. *Convection.* A carrier gas (usually air) supplies the heat for the evaporation of the liquid, i.e., the conversion of sensible heat into latent heat. The carrier gas subsequently removes the volatile matter.
2. *Conduction.* The heat is supplied indirectly and the carrier gas serves only to remove the evaporated liquid. Typically, the air quantity is approximately 10% of the quantity used in a convective process.
3. *Radiation.* This type of drying can be nonpenetrating, such as the drying of paint by infrared (IR) radiation, or penetrating, by microwaves. For a material to be effectively evaporated by microwaves the evaporated molecule must have a dipole. Microwave drying is the only process in which heat is developed in the material being dried rather than having heat diffused into the material. Again a carrier gas is required to remove the evaporated liquid.

Usually, a combination of two or more mechanisms is encountered in many dryers.

A distinction should be made between free and bound water. Initially, free water is evaporated until the critical moisture content is reached. The free water's latent heat during evaporation is essentially equal to that on evaporating from a pool, with the heat transfer being the rate-determining step. The evaporation occurs at a constant rate if the heat supply is constant. Drying to below the critical moisture content requires the evaporation of bound water, with the evaporation rate decreasing if the heat supply is kept constant. Bound water can be located in pores or crevices, can be physically absorbed, or can be present as water of hydration. The latent heat of evaporation of bound water is usually higher than that of free water; e.g., the ratio of the latent heats of evaporation of water on wool containing by weight 16% and 30% water (the critical moisture content) is approximately 1.1:1.

In Chapter 2, it is recommended that the drying step not be considered in isolation but rather be reviewed in the context of the entire process. Upstream process modifications can have a large impact on the drying stage, whereas the method of drying is often of paramount importance to product quality.

Unlike with, e.g., a centrifuge, usually one refers to a dryer as consisting of a number of pieces of equipment grouped together in subsystems. It is therefore more correct to refer then to drying systems. Convective drying systems are more complex than contact or radiation dryer systems. Drying is often the last processing step, which is followed by a solids-handling system, traditionally dealt with by mechanical engineers. In addition, being an energy-intensive process, drying may often be handled by energy specialists. It can therefore be considered a unit operation that falls at the interface of three disciplines, namely, chemical, mechanical, and energy engineering.

Procedures for determining the optimum dryer are covered in Chapter 3. One scheme is presented for continuous dryers with a separate scheme for batch dryers. Chapter 4 provides an introduction to convective drying and Chapters 5 through 8 cover in detail the four main categories of convective dryers. In these chapters, the performance of dryers has been analyzed, their literature data interpreted, and the design methods covered. The material that is presented permits an estimation of both fixed and variable costs for convective dryers. For this operation, computer programs are included in the appendixes.

In Chapter 9, miscellaneous convective and conductive continuous dryers are discussed and batch dryers are treated in Chapter 10. Special drying techniques, such as infrared and microwave drying, are dealt with in Chapter 11, and the very important issue of safety is covered in Chapter 12. Gas and dust explosions are also considered in that chapter. Chapter 13 covers continuous solids- and gas-moisture measurement, dryer control, and energy recovery. The separation of particulate solid material from spent drying gas by means of cyclones, fabric filters, and scrubbers are the topics in Chapter 14, and the selection of feeders for dryers is taken up in Chapter 15.

Finally, it should be stressed that drying is an expensive means of accomplishing a liquid-solid separation; e.g., in a convective dryer, 2–3 kg of steam is required for the evaporation of each kg of water. Performing a solid-liquid separation by means of a centrifuge or filter is usually much cheaper than using a dryer.

2

DRYING AS PART OF THE OVERALL PROCESS

When in the early stages of investigating a drying problem, attention should be given to the entire manufacturing process. This holistic approach may yield one of the following conclusions.

1. The dried product should have a certain residual moisture content.
2. The drying step is simplified via a process change.
3. A combination of a drying step with one or more process steps is implemented.
4. The water is removed by a nonthermal method.
5. The drying step is avoided by changing the process.
6. The product is not dried; however the process is not changed.

These six options will now be examined in more detail.

2.1   Residual moisture

To dry a product to a moisture content of zero often requires a great deal of energy but it is sometimes sufficient to dry a product to a specific moisture content before selling it. This procedure reduces energy costs and it is advantageous that more product be sold at the same raw material cost. This option can be useful in combination with a reliable in-plant continuous moisture-monitoring system.

2.2   Process change to simplify drying

Drying can often be simplified by increasing the particle size in the dryer feed. Various techniques, which are briefly covered below, can be used for particle-size enlargement. More detailed information can be found in standard textbooks on crystallization and precipitation (e.g., Mullin, 1971).

Solubility of the material in the solvent affects the particle size. Materials having moderate solubility in the solvent system being utilized (i.e., 1–30%) are generally obtained in a coarse form with a weight average particle size of 0.2–2 mm. This finding can be explained qualitatively since a small supersaturation/solubility ratio tends to lead to large crystals. For example, this behavior is found in sodium chloride, potassium chloride, and sugar.

Materials having a solubility of about less than 0.1% by weight tend to be obtained as small particles; e.g., gypsum has a weight average particle size in the range 1–100 μm.

Particles that are of size range 0.2–2 mm generally contain 1–5% moisture by weight when entering a dryer, whereas smaller particles may retain up to 30–40% by weight when discharged from a filter or centrifuge.

Particle size can be increased by changing the solubility of the dissolved material, by changing the solvent or pH, or by increasing the temperature, slurry density, or residence time. The average particle size usually decreases on increasing the system velocity (e.g., at the pump tip and in the heater tubes) or the supersaturation. Combinations of more than one of the above parameters can also be used to achieve a desired particle-size distribution.

Seeding the crystallizer contents can also increase the particle size. This procedure is applicable to systems that do not nucleate readily because of high viscosity, for example. Up to a certain level, supersaturation increases at which point many nuclei may be produced. Seeding is practiced to prevent this, in, e.g., sugar crystallization.

A particle-size decrease can be achieved by seeding a crystallizer containing a material that readily nucleates, e.g., sodium chloride. Sometimes, because of product specification, it is not desirable to alter the average particle size; e.g., rapid dissolution or proper dispersion of the product may require a small particle size. The average particle size and particle-size distribution are not the only factors that influence the moisture-retaining properties of materials; the particle shape (habit) and thus the specific area can also have a significant influence.

It is also possible to influence the process at the point where the particulate material is formed by replacing conventional crystallization (precipitation), liquid-solid separation, and drying equipment by drum-drying or spray-drying systems. Williams-Gardner (1971) gave two possible routes for the processing of clay tile-body suspension in water: (a) filter press—dryer—granulating unit—tile presses and (b) spray dryer—tile presses. Spray drying and drum drying do not lead to the problem of disposing of an impure mother liquor.

Another example of process change comprises leaching of the cake in a liquid/solid separation system at an elevated temperature, which causes a reduction in the viscosity of the adhering liquid and a more efficient dewatering step. This goal can be achieved by the use of steam in a leaching stage. Váhl (1957) described a dramatic effect in the sugar industry: (a) leaching with cold water yields a sugar cake at 40°C containing ca. 2% water by weight; (b) treatment with steam results in a sugar cake at 80°C containing about 0.6% water by weight with the additional benefit that further water loss occurs on the way to the dryer so that the cake arrives at the dryer containing only 0.2–0.3% water by weight.

In an article entitled "Steam dewatering of filter cakes," Simons and Dahlstrom (1966) reported moisture reductions exceeding 60% for permeable filter cakes (e.g., a crystalline heavy inorganic chemical with 50% by weight >200 μm and a narrow size distribution). However, impermeable filter cakes cannot readily be dewatered further.

The product as produced in the crystallizer/precipitator can be accepted, and an additional step can be introduced prior to liquid/solid separation and drying. Solids having a melting point of less than 100°C can be liquefied by the injection of live steam, i.e., the system is changed from being a liquid-solid to a liquid-liquid one. Subsequent cooling will lead to solidification and, if carried out correctly, can result in a particle-size increase. The process is termed *granulation.*

In suspension agglomeration, a binder liquid that is immiscible with the suspending liquid and that preferentially wets the solid surface is added to the slurry (on occasions it is necessary to add an auxiliary component to perform the wetting function). On mixing, the binder liquid gradually displaces the suspending liquid from the solid surface and agglomeration proceeds, resulting in a solid having a particle size of between one and several millimeters. It is possible to achieve suspension agglomeration of only one component in a multicomponent system by careful selection of the binder and wetting agent.

Zuiderweg and Van Lookeren Campagne (1968) described the recovery of 1–3% soot by weight (5–10 μm) from an aqueous suspension by means of oils (see Figs. 2.1 and 2.2). Because of the large surface area of the soot (e.g., 1000–1500 m^2/g for soot from heavy fuel oil), oil-soot ratios of 3–5 are required. An agitator power input of 15 HP/m^3 was used in a continuous pilot plant to obtain 2–7 mm pellets.

Bemer (1979) studied small-scale batch agglomeration using glass and aluminum silicate powders suspended in carbon tetrachloride with mixtures of water and glycerol acting as binder liquids. Continuous agglomeration of chalk (calcium carbonate) suspended in water, using a mixture of kerosene and oleic acid (oleic acid converting the hydrophilic surface of the chalk into a hydrophobic one) was also investigated. The latter set of materials was also used by Bos (1983). Mixing appeared to be an important

Figure 2.1 A number of continuous agglomerators to recover soot from an aqueous suspension. (Courtesy of Babcock Duiker, Kwintsheul, The Netherlands.)

Figure 2.2 A scale model of a continuous agglomeration plant to recover soot from an aqueous suspension. (Courtesy of Babcock Duiker, Kwintsheul, The Netherlands.)

factor. The agglomerated products contained from one to several tens of percentages by weight of the bridging agent. Capes (1980) reviewed this technique.

2.3 Combination of drying and other process steps

Since many possibilities exist in this category, drying can thus be combined with a chemical reaction, evaporation, mechanical liquid/solid separation, particle-size enlargement, and several other operations.

Example 1

The Solvay process is widely used for the manufacture of sodium carbonate (soda). Sodium bicarbonate is an intermediate particulate product separated from the mother liquor by rotary vacuum filters in which leaching of the crystals with water is also performed. Centrifuges are also used. The cake from the filters contains about 14% water by weight whereas from the centrifuges it typically contains about 8%. Calcining of the bicarbonate to soda and drying takes place in a single indirectly heated rotary-drum calciner, with the drying preceding the calcining.

$$2NaHCO_3 \rightarrow CO_2\uparrow + H_2O\uparrow + Na_2CO_3$$

Carbon dioxide is recycled after compression, and steam is usually used as heating medium. The hot soda ash (ca. 200°C) is cooled, screened, and packed-out or shipped in bulk. The product is called *light ash* because of its low bulk density.

Example 2

During the manufacture of potato chips, potatoes are peeled, sliced, and then washed in 60°C water. The wet slices are then added to 160°C oil, and curing and drying occur in one step. Salt and spices are added before packaging.

Example 3

It is possible to combine liquid/solid separation, leaching, and drying in a single unit that functions batchwise. The slurry is pumped to a filter-dryer and the dry solids are discharged at the end of the cycle. This setup is used if, for example, it is desirable to protect the operators from dust (see Figs. 2.3 and 2.4). Basically, the equipment comprises a closed Nutsche-type vacuum filter with options. A typical cycle is made up of: (a) feeding the slurry to the filter, (b) mother liquor removal, (c) leaching (displacement

Figure 2.3   A multifunction closed Nutsche vacuum filter. (Courtesy of Rosenmund, Liestal, Switzerland.)

or reslurry), (d) smoothing and compressing the cake to remove liquid, (e) drying (indirect heat transfer, moisture-entraining gas), and (f) discharge. Possible applications of this procedure include automation of existing processes, processing light-sensitive materials, solvent recovery, and handling toxic materials.

Reference was made in Section 2.2 to spray drying and drum drying. With the use of this equipment, evaporation, crystallization, liquid/solid separation, and drying can be combined. The combination is also possible in a thin-film evaporator (see Fig. 2.5). The choice of such a system may not be energetically preferable; however, it can be the best choice if conventional crystallization, liquid/solid separation, and drying are complicated or impossible or if the resulting product has favorable properties (e.g., the rapid dissolution of spray-dried coffee powder in which the formed particles appear as hollow spheres).

2.4   Nonthermal drying

As a general rule, moisture that can be removed mechanically should not be removed thermally. In this section, the mechanical removal of moisture is not considered; but physical absorption and chemical reaction, the two principal nonthermal methods for moisture removal, are.

Example 1

An inorganic hydrate leaves a centrifuge containing a small percentage of moisture. The material can be admixed in a screw conveyor with a small amount of a lower hydrate that picks up the water and is itself converted to the higher hydrate.

$$X \cdot aH_2O + bH_2O \rightarrow X \cdot (a + b)H_2O$$

Example 2

In casting operations for certain pottery objects, as, e.g., for teapots, gypsum molds are applied to dewater a clay stream (slip) in order to produce an article in the correct physical form. The slip typically contains 30–40% water by weight. Additives allow the concentrated slurry to flow. A porous gypsum mold absorbs the water and must be dried before it can be reused. After 10 to 20 cycles, the pores of the mold are plugged and it cannot be regenerated further.

Example 3

Williams-Gardner (1971) mentioned the use of starch molds for the shaping of confections. Some of the water in the syrup is absorbed to accomplish product shaping.

Example 4

The shaping of powdery drugs into granules can be carried out in a single vessel. This process involves drying and is required because powders are too cohesive to be fed directly into a tabletting machine. The granulation stage is usually carried out using a binder liquid (water). The drug ingredients are mixed with water into which they are soluble to some extent, thus creating the granulation. The granules are subsequently dried and the crystal bridges confer strength. Fielder (of Southampton, England) markets

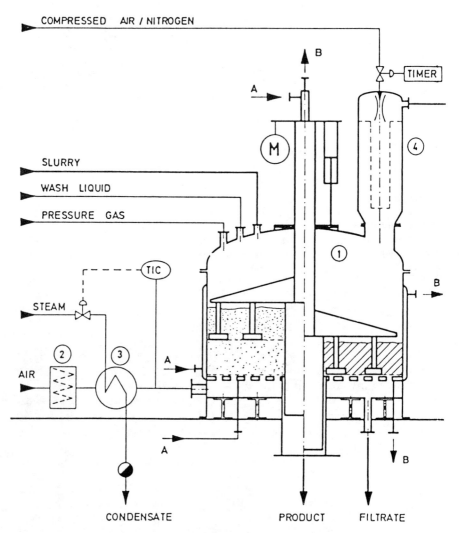

Figure 2.4 A process-flow diagram showing the operation of a Rosenmund Filter-Dryer. (Courtesy of Rosenmund, Liestal, Switzerland.)

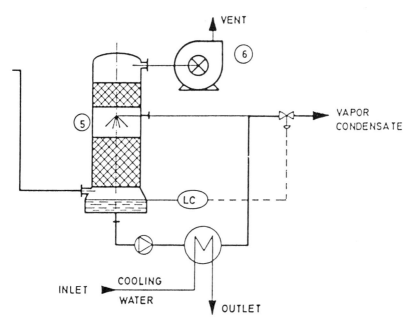

VENT

VAPOR
CONDENSATE

LC

COOLING
WATER

INLET

OUTLET

① ROSENMUND FILTER
② INLET AIR FILTER
③ AIR HEATER
④ DUST FILTER
⑤ SCRAPER
⑥ BLOWER
A WARM WATER INLET
B WARM WATER OUTLET

Figure 2.5 A thin-film evaporator in the manufacturer's plant. (Courtesy of Buss, Basel, Switzerland.)

equipment in which mixing, granulation, and drying by means of vacuum microwave heating takes place. This type of drying leaves the granules at rest, which does not occur with fluid-bed drying where the particles are carried to and even through filter bags. Losses such as these can be expensive and perhaps prohibited because of environmental considerations. Microwave drying is better than tray drying because it proceeds faster and can be combined with mixing and granulation.

2.5 Process changes to avoid drying

In this instance, process change enables the product to be obtained in a form that makes drying superfluous.

Example 1

Acetylene is manufactured by the reaction of calcium carbide and water:

$$CaC_2 + 2H_2O \rightarrow Ca(OH)_2 + C_2H_2 \uparrow \qquad -134 \text{ kJ/mol}$$

Originally, an excess of water was used to control the reaction and to suspend the lime, the hydrated lime slurry being sent to a second plant for use there. Because of changes in the latter plant, the practice of pumping the slurry had to be discontinued. Liquid/solid separation and drying of the lime was considered; however, a better solution was to switch from a wet to a dry manufacturing process. This process uses about 1 kg of water per kg of carbide and the heat of reaction is dissipated by the vaporization of the water. The hydrated lime still contains about 5% water by weight but can be directly packed and sold to the building trade, for example. The rates of water addition and mixing are critical and must be carefully controlled, nevertheless the process is now used widely.

Example 2

Van Holten and Ribbens (1971a,b) developed a method to circumvent the drying of aliphatic di-acyl peroxides, which are solid at room temperature and fusible below 120°C. These compounds are manufactured by the addition of an acid chloride to an aqueous solution of sodium peroxide.

$$2R - \overset{\displaystyle O}{\overset{\|}{C}} - Cl + Na_2O_2 \rightarrow R - \overset{\displaystyle O}{\overset{\|}{C}} - O - O - \overset{\displaystyle O}{\overset{\|}{C}} - R + 2NaCl$$

The reaction proceeds at room temperature, with the chemical reaction being followed by precipitation. Since the products are almost insoluble in the aqueous phase, the resultant particles are small. Conventionally, the solid is separated from the liquor, then leached and dried in plate dryers. The alternative process route is to inject live steam into the reaction slurry as it is pumped through a line, the system temperature being elevated to a value higher than the melting point of the peroxide. The liquid-solid system is converted into a liquid-liquid system that can be separated in a disk centrifuge. The organic phase is rapidly solidified on a cooling belt. The peroxide is held at the elevated temperature for a short period of time to avoid decomposition. Although the short process time is an important advantage, the principal benefit is that the product which is obtained is in a purer form than when obtained by the conventional process, because leaching of the filter cake is very difficult. Solid-handling characteristics are also improved with the product being obtained as a flake rather than as a powder.

Example 3

A similar process improvement has been developed by Appel and Brossmann (1984) for the manufacture of dialkyl peroxy-di-carbonates. An ester of chloroformic acid reacts with an aqueous solution of sodium peroxide at an ambient or slightly elevated temperature:

$$2R-O-\overset{\overset{\displaystyle O}{\|}}{C}-Cl + Na_2O_2 \rightarrow R-O-\overset{\overset{\displaystyle O}{\|}}{C}-O-O-\overset{\overset{\displaystyle O}{\|}}{C}-O-R + 2NaCl$$

(where R is a C6–C18 linear, cyclic or branched alkyl group). It is also possible to use a mixture of esters. The patent principally concerns the manufacture of both dimyristyl- and dicetyl-peroxy-di-carbonate. Once again live steam is injected into the reaction slurry to liquefy the product, which is separated from the aqueous phase by means of a disk centrifuge. The product is subsequently solidified using a flaker. The particulate product contains little dust.

2.6   No drying

The process is left as is except for the drying stage, which is simply not carried out. On analyzing the full picture, it sometimes becomes evident that the manufacturer of a product takes pains to dry the product knowing full well that upon its receipt the customer dissolves or suspends the particles in water. If, for example, packaging, transporting, and unloading of the product is not hindered, a delivery of a wet cake is simpler. In general, this can be realized for materials that are not soluble in the adhering liquid because soluble products may present caking problems.

Example 1

Dibenzoyl peroxide is a solid product prepared by reaction/crystallization in an aqueous phase. It was originally marketed as a dry powder obtained through liquid/solid separation followed by drying. But in this form the dry powder has unfavorable safety characteristics (low impact resistance) and severe in-plant decompositions have been experienced during its manufac-

Figure 2.6 Centrifuging and drying can be combined in this piece of equipment. (Courtesy of Gala Industries, South Charleston, West Virginia.)

ture. A re-evaluation of the market led to customer acceptance of a wet-cake form and mixtures of the dry product and inert fillers (phlegmatization).

Example 2

High-pressure polyethylene is extruded and cut into pellets in water. The water-pellet slurry can be transported to a dewatering screen. The wet pellets are subjected to centrifugal action and they can also be dried with warm air simultaneously and then bagged for sale (see Fig. 2.6)

References

Appel, H. and G. Brossmann (1984). Process for the continuous manufacture of dialkyl peroxy-di-carbonates. European Patent 0 049 740 (in German).

Bemer, G. C. (1979). *Agglomeration in Suspension: A Study of Mechanisms and Kinetics,* Delft University of Technology, Delft, The Netherlands. Ph.D. Thesis.

Bos, A. S. (1983). *Agglomeration in Suspension,* Delft University of Technology, Delft, The Netherlands. Ph.D. Thesis.

Capes, C. E. (1980). *Particle Size Enlargement,* Elsevier Scientific Publishing, Amsterdam, p. 161.

Mullin, J. W. (1971). *Crystallisation,* Butterworths, London.

Simons, C. S. and D. A. Dahlstrom (1966). Steam dewatering of filter cakes, *Chem. Eng. Progr.,* 62, 75.

Váhl, L. (1957). The drying of particulate material, elucidated with sugar and starch as examples, *De Ingenieur* 69, 77 (in German).

Van Holten, J. and C. Ribbens (1971a). Improvements in or relating to the purification of organic peroxides. British Patent 1 239 088.

Van Holten, J. and C. Ribbens (1971b). Apparatus useful in the purification of organic peroxides. British Patent 1 239 089.

Williams-Gardner, A. (1971). *Industrial Drying,* Leonard Hill, London, p. 5.

Zuiderweg, F. J. and N. van Lookeren Campagne (1968). Pelletizing of soot in waste water of oil gasification plants—the Shell pelletizing separator (S.P.S.), *The Chem. Eng.* (London) No. 220, CE 223.

3

PROCEDURES FOR CHOOSING A DRYER

3.1 Introduction

A large variety of drying equipment is currently available from manufacturers. In this chapter, the screening procedures that offer a preliminary choice for a specific drying duty will be described. Subsequent chapters in this book provide more details for the main classes of dryers, e.g., dimensions, capital cost, and energy consumption.

The selection schemes that will be described here are for batchwise and for continuous dryers. They do not cover every possible type of dryer but most of the industrially important systems are considered. The correlation between dryer size and investment cost was discussed by Noden (1969). By updating this approach and factoring energy consumption, an estimate can be made of the drying cost for various dryer types.

Production capacities exceeding 100 kg/hr often require a continuous dryer, but the choice between batchwise and continuous dryers also depends upon the nature of the equipment preceding and following the dryer. Table 3.1 outlines the data that have to be collected before selection of a dryer system can be started, and Table 3.2 lists some of the criteria for evaluating dried particulate material.

Steam-tube and direct-heat rotary dryers are universally applicable. They can be chosen if there is a limited amount of experimental data or

Adapted and reprinted by special permission from *Chemical Engineering,* March 5, 1984. Copyright © 1984, by McGraw-Hill, New York, NY 10020.

19

Table 3.1   Data to Be Assessed Before Attempting Dryer Selection

Production capacity (kg/hr)
Initial moisture content
Particle-size distribution
Drying curve
Maximum allowable product temperature
Explosion characteristics (vapor/air and dust/air)
Toxicological properties
Experience already gained
Moisture isotherms
Contamination by the drying gas
Corrosion aspects
Physical data of the relevant materials

(Methods for determining the numerical values of the various criteria must be agreed upon.)

Table 3.2   Some Criteria for Judging a Dried Particulate Material

Moisture content
Particle-size distribution
Bulk density
Hardness
Dust content
Flow characteristics
Color
Odor, taste
Appearance
Dispersibility
Dissolution or rewetting behavior
Assay
Caking tendency
Segregation of originally dissolved components (food)

(Methods for determining the numerical values of the various criteria must be agreed upon.)

insufficient time to go through the selection procedures described in this chapter. However, the tailor-made solution, using the selection procedure, will often lead to a less expensive investment in solving the drying problem.

3.2 Selection schemes

Figures 3.1 and 3.2 provide step-by-step procedures for the selection of a batch dryer and a continuous dryer, respectively, and the information present in each section supplements the respective chart. Batch dryers will be discussed first.

Vacuum dryers

If the maximum product temperature is lower than or equal to 30°C, it is worthwhile to look at a vacuum dryer. A good driving force for evaporation can be created while keeping the temperature low. The vacuum tray dryer is the simplest, but the product must usually be sieved to break down any agglomerates (the breakdown may be aided mechanically).

The capacity of the vacuum tray dryer is rather low. It may be economic to consider an agitated vacuum dryer (Fig. 3.3) in which the contents are moved mechanically. Such dryers are widely used.

If the product is oxidized by air during drying, consider either vacuum drying or inert-gas drying.

If either the product or the removed liquid is toxic, the equipment must be kept closed as much as possible. Again, a vacuum dryer can render good service. (In addition, dust formation is avoided.)

Fluidized-bed dryers

If the average particle size is about 0.1 mm, or larger, fluidized-bed drying (Fig. 3.4) may be considered. (If smaller particles must be dealt with, the equipment required to handle them may be too large to be feasible.) Inert gas may be used if there is the possibility of explosion of either the vapor or dust in air.

If such a dryer is being considered, it is easy to carry out tests in a small fluid-bed dryer.

Other dryers

As Figure 3.1 shows, the remaining possibilities regarding batch drying are the tray dryer and the agitated pan dryer.

Solvent evaporation

In continuous drying, if a solvent must be evaporated and then recovered, it is usually not optimum to choose a convection dryer. Since solvent must be condensed from a large carrier-gas flow, the condenser and other equipment become rather large.

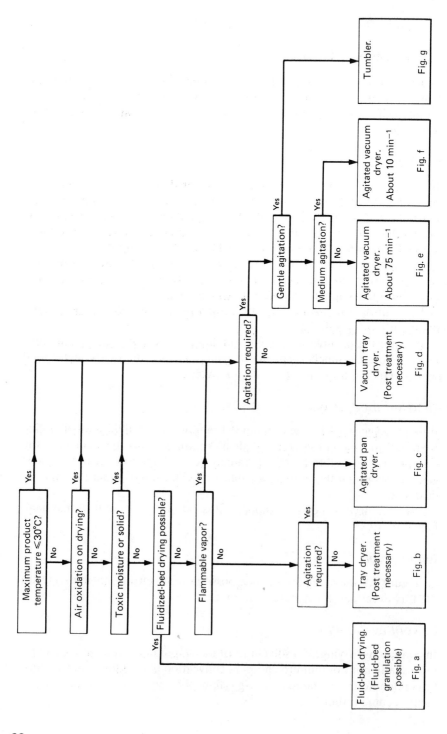

22

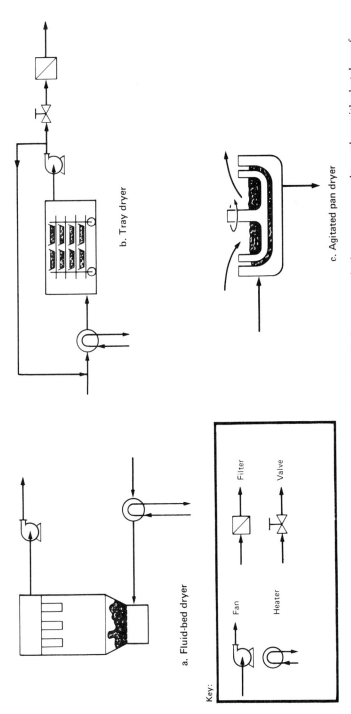

Figure 3.1 Decision tree for the selection of a batch dryer suitable for any particular process need together with sketches of the various dryers suggested. (Continues on next page.)

a. Fluid-bed dryer

b. Tray dryer

c. Agitated pan dryer

Key:

Fan

Heater

Filter

Valve

23

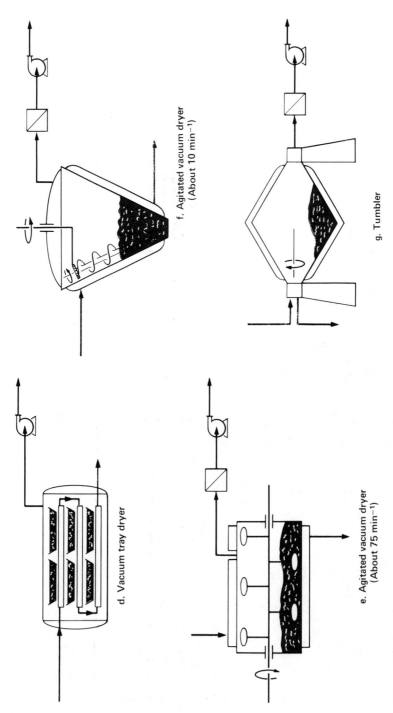

d. Vacuum tray dryer

e. Agitated vacuum dryer
(About 75 min⁻¹)

f. Agitated vacuum dryer
(About 10 min⁻¹)

g. Tumbler

Figure 3.1 Continued.

24

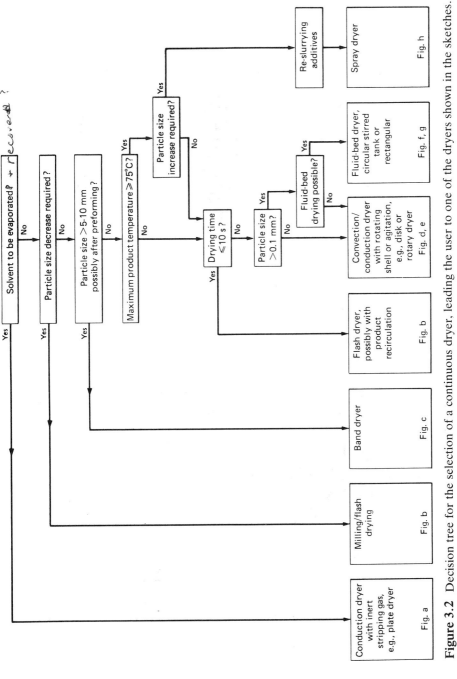

Figure 3.2 Decision tree for the selection of a continuous dryer, leading the user to one of the dryers shown in the sketches. (Continues on next page.)

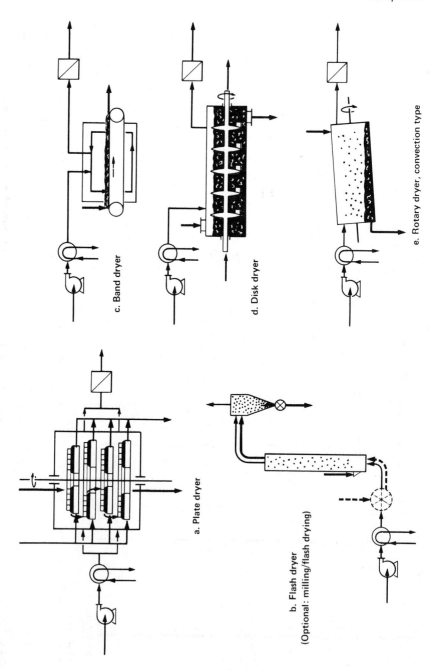

c. Band dryer

d. Disk dryer

e. Rotary dryer, convection type

a. Plate dryer

b. Flash dryer
(Optional: milling/flash drying)

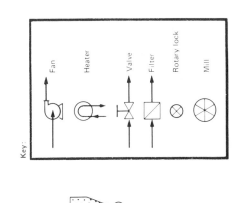

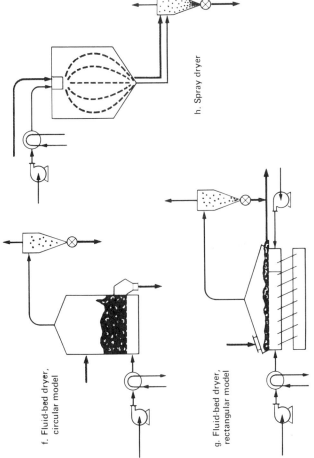

f. Fluid-bed dryer, circular model

g. Fluid-bed dryer, rectangular model

h. Spray dryer

Figure 3.2 Continued.

Figure 3.3   Agitated vacuum dryer with solvent recovery. (Courtesy of Hosakawa-Nauta, Haarlem, The Netherlands.)

Milling/drying

If it is necessary to decrease particle size in addition to drying, the two operations may be advantageously combined. The wet particulate solid is transported by warm or hot gas into a mill. Gas and particulate solid leave the mill, fly through a line, and are separated. The comminution often greatly helps the drying by exposing internal moisture. This type of drying is encountered in cases where the fineness is of great importance to the application. Examples are cases where a rapid and complete dispersion (or dissolution) or a high activity (m^2/g) are being aimed for.

Figure 3.4   Batch type of fluid-bed dryer. (Courtesy of Glatt, Binzen, Federal Republic of Germany.)

Band (belt) dryers

A band dryer (Fig. 3.5) is preferable if the particles are rather coarse (i.e., over 5 to 10 mm). The particles are evenly spread onto a slowly moving, e.g., 5 mm/sec, perforated belt. The belt moves into a drying cabinet and warm gas passes downward through the layer. This type of dryer is chosen when it is not possible to suspend the particles in the drying gas. The dryer must offer a residence time, say 15 min, because bound moisture must diffuse through the pellet.

The performance of such a dryer can be predicted from the determination of the drying curve on a small scale, employing realistic conditions

Figure 3.5 Band dryer uses a slowly moving perforated belt. (Courtesy of Babock-BSH, Krefeld-Uerdingen, Federal Republic of Germany.)

(pellet characteristics, layer thickness, and drying-gas parameters). Many band dryers are used to dry preformed particles. The wet particulate material is mixed with additives, granulated, and dried. One reason for doing this is that direct drying of the wet material may yield a dusty material, whereas the granules are less dusty. Band dryers are also used in food applications, e.g., diced carrots.

Spray dryers

A spray dryer (Fig. 3.6) can be used if the aim is the conversion of a fine material (e.g., 15 μm) into a coarser material of spherical form (e.g., 150 μm). The material thus obtained is free-flowing and less dusty. However, this is an expensive drying method.

The wet filter cake is reslurried (for example, to 40% solids by weight), additives are introduced, and the mixture is fed to the spray dryer, where the liquid is evaporated. To keep the size of the equipment reasonable, a minimum inlet-gas temperature is required (perhaps 200°C) to produce a solids outlet temperature that exceeds 75°C.

Flash dryers

The flash dryer is the workhorse of industry (Fig 3.7). However, because drying must take place within 10 sec, the removal of bound moisture is

Figure 3.6 Spray dryer produces a spherical-shaped product. (Courtesy of Niro Atomizer, Soeborg, Denmark.)

Figure 3.7 Flash dryer—one of the most common types used in industry. (Courtesy of Babock-BSH, Krefeld-Uerdingen, Federal Republic of Germany.)

difficult. Since the dryer is essentially a vertical line, drying and vertical transport can be combined.

Fluid-bed dryers

Use of a fluid-bed dryer is a possibility if the particle size exceeds 0.1 mm. A round piece of equipment holding a thick product layer is one option. The holdup must be large, as the composition of the dryer contents equals the outlet composition. The thick product layer means that much fan power is needed to push the drying gas through it. Because caking will not easily occur, the construction can be stationary. Such constructions can allow high drying-gas temperatures—up to 500–600°C.

A rectangular-shaped dryer will permit plugflow. Fig. 3.8 shows a stationary type with a thick product layer; a shallow layer may require vibration for transport and to prevent caking. However, a vibrated construction (Fig. 3.9) cannot withstand high temperatures. A realistic maximum drying-gas inlet temperature is 300°C. Moreover, the hot drying gas must pass through the flexible devices that couple moving and stationary parts; generally, these cannot withstand high temperatures either.

Miscellaneous dryers

Jobs that cannot be handled by the fluid-bed dryer or flash dryer can often be accomplished in a conduction dryer, such as a steam-tube rotary dryer (for dusty products), or in a convection dryer, such as a rotary dryer.

Figure 3.8 Continuous, stationary fluid-bed dryer. (Courtesy of Sulzer-Escher Wyss, Ravensburg, Federal Republic of Germany.)

Figure 3.9 Continuous, vibrated fluid-bed dryer. (Courtesy of Sulzer-Escher Wyss, Ravensburg, Federal Republic of Germany.)

Powerful combined convection/conduction dryers also fall under this heading.

3.3   Drying liquids, slurries, and pastes

Up to this point we have mainly been considering the drying of masses of wet particulates, such as filter cakes. Here are some things to consider when choosing equipment for continuous drying of liquids, slurries, and pastes:

Spray dryers

These may be chosen if the isolation of a solid from a solution or slurry via conventional crystallization and liquid/solid separation is either impossible or too complicated. Sometimes, spray drying is chosen if the characteristic spherical particle shape is desired (as in making instant coffee). Typically, the average particle size (on a weight basis) falls in the range of 50 to 200 μm.

This dryer's short residence time is a plus for heat-sensitive products, e.g., milk powder and organic salts.

Film drum dryers

As with spray dryers, drum dryers may be chosen in cases where crystallization and liquid-solid separation are not feasible. This type of dryer can be used for pastes, and it can be placed under vacuum for heat-sensitive products. Examples of its use include the flaking of mashed potatoes and the drying of other instant foods.

Cylindrical scraped-surface evaporator/crystallizer/dryers

These dryers resemble the equipment used for other thin-film techniques. The fluid to be dried passes through in plugflow. Between the feed point (for a solution or suspension) and the product outlet (for a more or less free-flowing powder), there is a zone where much power is consumed (per unit wall area), as a viscous paste is converted into particles. In principle, the equipment is suitable for pastes as well as liquids and slurries. Vacuum may be applied for heat-sensitive products, and solvent recovery is possible.

Agitated pan dryers or agitated vacuum dryers (Fig. 3.10)

These should be considered for batchwise drying of liquids, slurries, or pastes.

In both of these dryers, solvent recovery is possible. In the case of liquids and slurries, as the drying proceeds, the power consumption rises to a maximum and then decreases. (As noted above, a liquid or slurry is converted into a more or less free-flowing powder via a viscous paste.) Hence, the motor, gearing, and stirrer must be adequately sized for the maximum power consumption.

Figure 3.10 Agitated vacuum dryer for batch drying of liquids, slurries, and pastes. (Courtesy of Babock-BSH, Krefeld-Uerdingen, Federal Republic of Germany.)

3.4   Special drying techniques

Infrared drying

This is generally confined to surface drying, as of newly painted automobile bodies.

Freeze drying

In this technique, ice is sublimed from a frozen product while under vacuum. Drying temperatures typically range from -30 to $-10°C$, so the process is ideal for very heat-sensitive products, such as those of a biological nature.

Microwave drying

This process is used for relatively large and expensive articles that contain little moisture. It has also recently been applied to the drying of pharmaceutical products under vacuum, and the drying of pastas (Hubble, 1982) to avoid case-hardening.

3.5   Some additional comments

Conduction vs. convection dryers

For both batch and continuous dryers, a distinction can be made between conduction and convection drying. Convection dryers (e.g., flash and spray dryers) often require relatively large solid/gas separation equipment. Thermal efficiency of convection dryers increases with increasing inlet temperature of the drying gas. There is no such effect with conduction dryers. Furthermore, conduction dryers may be preferred for dusty products.

A different aspect is the relationship between product temperature and heating-medium temperature. In convection dryers, the product usually adopts the adiabatic saturation temperature. But in conduction dryers, product in contact with hot metal will attain the metal's temperature.

Finally, the residence time in conduction dryers can be up to several hours; but the residence time in flash and spray dryers can be short—ten seconds or less. Both time and temperature are important in regard to thermal degradation.

In the classification we have used, the plate dryer (Fig. 3.11) is a hybrid type. This type of dryer can be useful for solvent recovery or dusty products. The raking action counteracts caking.

Figure 3.11 Plate dryer is useful for handling dusty products. (Courtesy of Krauss Maffei, Munich, Federal Republic of Germany.)

Combining dryer types

A combination of two different types of dryers is sometimes optimum. One example is the spray drying of milk, followed by a posttreatment in a fluid-bed dryer.

Dust explosions

If there is a possibility of a dust explosion in a convection dryer, special precautions must be taken. These may involve use of inert gas, provision of explosion panels, triggered injection of suppression agents, and the like.

Scaling up

Some of the dryers thus far discussed have a limited possibility for scaling up. In comparing a drum dryer with a spray dryer, it is clear that the spray dryer can cope with larger evaporation loads.

There is practically no limit to scaling up a rotary dryer. Convection-type rotary dryers are well established in the mining industry.

Vacuum drying

Continuous vacuum drying is not often found; it is difficult to feed and extract the material under vacuum. However, there are some special cases where a liquid, slurry, or pumpable paste is fed and a dried solid (with relatively good handling characteristics) is removed.

3.6 Testing on small-scale dryers

For testing, it is usually wise to seek the cooperation of a limited number of reputable drying-equipment manufacturers. It is often possible to ship samples of the material to be dried and have them tested in the manufacturer's small-scale units. This is relatively simple and straightforward, but there are risks:

1. The material may have changed in character because of chemical or physical changes between the time the wet product was tapped off the process stream and the time of drying.
2. Because the quantity of sample shipped is limited, it is not possible to check long-run performance. For some desired process results, such as proper transport of the material through the dryer, and absence of caking or dusting, it is essential to carry out experiments over a long period.
3. The performance of the dryer is not checked in relationship to the process-plant infrastructure—for example, the behavior of the dried material in the plant's solids-handling equipment.

Consequently, if the experiments done by the equipment supplier are successful, it is good practice to install a small-scale dryer in the process plant, where it can be fed from a side stream. If a new product or process is being dealt with, the dryer should be in the pilot plant. If there is no pilot plant, representative material must be made or obtained and then dried at the equipment manufacturer's facility. If you are interested in seeing full-scale equipment in operation, the manufacturer may be able to arrange an introduction to a facility using such dryers.

Before purchase, it is necessary to obtain guarantees from the dryer manufacturer and to agree on the method of analysis that will determine whether the guarantees are met.

3.7 Examples of dryer selection

Example 1

In an existing plant, the rotary dryers being used for salt (NaCl) dated back to the 1930s. These continuously functioning dryers were worn out and needed replacement. After a successful drying test at a manufacturer's facility, it was decided to install a flash dryer. The choice was based on the following:

1. Water was to be evaporated.
2. There was no need to change particle size.
3. The average particle size was about 0.4 mm.
4. The inorganic salt was not temperature-sensitive.
5. Only surface moisture was to be removed.

The flash dryer was installed, but was not successful owing to the formation of a fine product mist from the equipment. Two reasons for this were suggested: (a) the high velocities in the flash dryer abrade the crystal, producing fine particles, and (b) the rapid evaporation of the surface moisture causes nucleation therein (the surface moisture being a saturated solution of salt in water).

The problem disappeared when the flash dryer was replaced by a troughlike vibrated fluid-bed dryer having a shallow product layer. Residence time is now minutes instead of seconds, and velocities are much lower.

Example 2

A plant was built for producing about 40 kg/hr (dry basis) of a solid organic peroxide. It was decided to install a batch dryer for these reasons:

1. Maximum product temperature was 40°C.
2. Water was to be evaporated (about 25% by weight).
3. The solid was not very toxic.
4. Oxidation by air did not occur.
5. A dust explosion was possible.
6. Average particle size was about 500 μm.

Testing established the feasibility of fluid-bed drying, using a recycled inert gas. The drying gas is warmed, indirectly, by warm water to 40°C. The diameter of the supporting plate is 0.92 m.

The process is controlled by measuring the temperature of the leaving gas. When it reaches a specified temperature, the warm water supply is switched off, and the batch is cooled by an inert gas for 10 min.

Example 3

A case of drying of organo-tin compounds is described by Schaake and Stigter (1975). Many of these compounds are toxic. Between 10 and 30% of moisture (a mixture of various solvents, which may include water) is to be evaporated. The maximum allowable product temperature is between 50 and 90°C. Vacuum used is between 30 and 50 Torr. Batch size is 1,500 to 2,000 kg, and the bulk density of the dry product is about 500 kg/m³.

Originally, a tumbler dryer was used, but it was later replaced (in order to increase capacity) by a conduction dryer comprising a conical mixer with a screwtype mixing element rotating along the circumference of the cone and around its own axis. Both dryers were of stainless steel. Their capacities were:

	Tumbler	Conical mixer
Drying time, hr	20–40	10–30
Volume, l	6,800	4,000

Hourly production is about 100 kg, so the choice of a batch dryer seems logical. A vacuum dryer was selected because of the toxicity of many of the products and the flammability of the vapors. The reasons for replacing the tumbler dryer by the vacuum dryer with medium agitation were:

1. Shorter drying time because of better heat transfer (no crusts).
2. Fixed provision for charging and discharging.
3. Homogeneous product—no manual cleaning because of the absence of crusts.
4. Low maintenance costs. No need to convey utilities from stationary parts to rotary parts (and vice versa) as in the tumbler.

According to the reference, the drying cost per unit of product was reduced by 40% by replacing the tumbler.

References

Hubble, P. E. (1982). Consider microwave drying, *Chem. Eng.*, 89, 125.

Noden, D. (1969). Industrial dryers—selection, sizing and costs, *Chem. Process Eng.*, 50, 67.

Schaake, P. and H. Stigter (1975). The vacuum drying of sensitive specialty chemicals, *J. Prep. Processing*, 16, 27 (in German).

4
CONTINUOUS CONVECTIVE DRYING

4.1 Introduction

Convective drying is accomplished by contacting a process stream with a warm- or a hot-gas stream. The heat required for evaporation of the moisture is provided by cooling of the gas stream; the gas stream is also used as a carrier for the removal of the evaporated moisture.

Convective dryers can be conveniently divided into three classes.

1. Dryers in which all the product is entrained by the gas stream (flash, spray, Convex)
2. Dryers in which part of the product is entrained (fluid bed, direct rotary, and Rapid)
3. Dryers in which a small but not negligible part of the product is entrained (conveyor, plate, tray)

The four convective dryers that are most important to industry will be discussed in Chapters 5–8, namely, continuous fluid bed, rotary, flash (pneumatic-conveyor), and spray dryers. Each chapter starts with a general description of the equipment and provides details of the suitability of the dryer for various drying functions. Theoretical aspects are also covered, namely, minimum fluidization velocity, literature design models for direct-heat rotary dryers, rapid drying of wet particles, and particle-size distribution on spray drying. The theoretical aspects are then followed by worked examples concerning dryer design. Computer programs in BASIC that allow calculation of variable costs and optimization of air-inlet temperature conclude these chapters. Information on dryer prices is supplied to permit cost estimates for different dryer sizes. Thus, Chapters 5–8 should provide convenient details for choosing among the four main types of convective dryers, whereas this chapter concentrates on aspects that are common for convective dryers.

4.2 Common aspects

Modes of flow

Four types of flow are identified:

1. Countercurrent
2. Cocurrent or parallel
3. A mixture of countercurrent and cocurrent
4. Cross

Countercurrent flow exposes the dried material to the highest temperature. This mode of flow permits a very low final moisture content but it also creates the risk of overheating the processed material. Parallel flow exposes the feed to the highest temperature, leading to a high initial drying rate. The fast initial drying basically occurs at the adiabatic saturation temperature; however, evaporation that is too rapid may damage the product.

Spray and rotary dryers can accommodate the first three modes, with the first two being the most frequently adopted. Flash dryers have parallel flow, whereas fluid-bed dryers have crossflow.

The flow mode does not strongly influence operation efficiency since the bulk of drying takes place when the solid is at the adiabatic saturation temperature.

Air-inlet and air-outlet temperatures

Air-inlet temperature varies between 100 and 800°C, whereas the corresponding air-outlet temperature varies only between 50 and 150°C. Generally, the outlet temperature of any continuous convective dryer is about 100°C.

Lower drying-gas temperatures, e.g., up to 200°C, are used to dry many organic chemicals that have temperature sensitivity. However, relatively low drying-gas temperatures (e.g., up to 300°C) are also used for coarse inorganic crystals that have between 1 and 5% of initial moisture by weight. If higher drying-gas temperatures are used, heating the dry solid flow consumes too much energy. Wet inorganic and stable organic materials can be dried with high air-inlet temperatures.

There is a relationship between the efficiency of the drying process and the air-inlet temperature, e.g.:

Case 1

TA_{IN} = 800°C TA_{OUT} = 150°C
Air at 20°C enters the burner
Dryer efficiency = (800 − 150)/(800 − 20) = 0.83

Case 2

TA_{IN} = 200°C TA_{OUT} = 75°C
Air at 20°C enters the burner
Dryer efficiency = (200 − 75)/(200 − 20) = 0.69

However, these ratios can be compared directly only if the heat absorbed by the solids can be neglected. This is because the solid will leave the dryer at a higher temperature in Case 1 than in Case 2.

The solid-outlet temperature will be lower than the air-outlet temperature for cocurrent flow. The difference will be greater for short residence-time dryers (flash and spray dryers). The solid-outlet temperature may exceed the air-outlet temperature for countercurrent flow and for crossflow in plugflow fluid-bed dryers.

Gas velocities

Table 4.1 gives the terminal velocities of spheres in air at 15, 90, and 205°C (Nonhebel and Moss, 1971). Appendix 4.1 contains a computer program for calculating terminal velocities of spherical particles in air (Coulson and Richardson, 1968). The examples are for a specific mass of 2,160 kg/m³.

Rotary dryers

The empty-tube drying-gas velocity is selected by considering the possible entrainment of solids. Table 4.1 helps the selection. Typical values are between 0.5 and 2.5 m/sec.

Flash dryers

The empty-tube drying-gas velocity is selected after considering the necessity to entrain all the solids. It is possible to mount a cage mill at the lowest point. Typical values are between 10 and 30 m/sec.

Fluid-bed dryers

The spent drying gas expands into the freeboard after passing through the product. The gas velocity in the disengaging height is selected so as to avoid excessive dust entrainment. Typical values are between 0.25 and 1 m/sec.

Spray dryers

The empty-tube gas-outlet velocities for spray dryers with rotary atomization are typically between 0.1 and 0.5 m/sec. For spray dryers with single-

Table 4.1 Terminal Velocities of Spherical Particles in Air

Diameter	$\Delta\rho = (\rho_P - \rho_F)$, kg/m^3					
	1,000			2,000		
	°C			°C		
μm	15	90	205	15	90	205
50	0.074	0.062	0.052	0.14	0.12	0.10
100	0.25	0.22	0.19	0.46	0.41	0.37
250	0.91	0.87	0.83	1.51	1.48	1.43
500	1.98	2.01	1.99	3.17	3.20	3.26
600	2.39	2.43	2.46	3.78	3.84	3.94
700	2.79	2.87	2.91	4.33	4.51	4.70
800	3.14	3.23	3.35	4.85	5.07	5.27
900	3.48	3.66	3.79	5.42	5.64	5.91
1,000	3.84	4.00	4.21	5.97	6.19	6.50
1,250	4.73	4.91	5.12	7.10	7.53	7.86
1,500	5.37	5.76	5.98	7.78	8.65	9.40
2,000	6.71	7.18	7.71	9.72	10.6	11.7
2,500	7.71	8.45	9.20	11.3	12.5	13.7
3,000	8.75	9.56	10.5	14.8	14.0	15.5
5,000	11.6	13.1	14.6	16.7	18.6	21.2

Source: Nonhebel and Moss (1971).

fluid nozzle atomization, 0.25 to 1 m/sec is normal. The lower range for rotary atomization is because of the relatively large diameter of the chambers with atomizing wheels.

Note well that the spent-gas outlet flow contains evaporated water; the water content being high for high-temperature flows and low for low-temperature flows. Table 4.1 should be used with care since it is valid for air only.

Method of heating

Heating can be direct or indirect. Direct heating involves the combustion of oil or gas and the passage of the combustion gases into the dryer. The composition is only slightly affected by the products of combustion because of the large mass of airflow. The drying-gas temperature can be as high as 1000°C. Incomplete combustion of oil or gas can lead to contamination of the product with soot. Direct heating of food products is infrequently performed because of the nitrosamine issue and the NO$_x$ issue.

Indirect heating can be carried out by means of steam or thermal oil. Steam heating is not normally used to provide temperatures above 200°C because of the high steam pressure required to produce high steam condensation temperatures, e.g., 16 bara steam condenses at about 200°C. Heat-exchange oil may be used to achieve temperatures to 300°C, with the oil being indirectly heated with a flame or by electricity. The use of heat-exchange oil is inherently safer than the use of direct heating.

Residence times

Flash and spray dryers are used to dry materials in seconds, whereas plugflow fluid-bed dryers with shallow (e.g., 0.1 m) product layers dry in minutes. Circular fluid-bed dryers with deep (e.g., 1 m) beds and rotary dryers normally have a residence time of between 15 min and 1 hr.

Heat losses

Two categories of heat loss can be distinguished: (a) losses during normal, steady-state operation and (b) additional losses that can be observed over relatively long periods (e.g., a year).

Losses during normal, steady-state operation

These losses can be calculated from performance data reported in the literature and as supplied from manufacturers of drying equipment (Perry et al., 1984 and Williams-Gardner, 1971). The calculated losses are in agreement with values obtained by plant measurements.

Causes of the losses are: (a) conduction, (b) convection, (c) radiation, and (d) ingress of air cooling the drying air or alternatively the loss of high-temperature drying air.

Factors influencing these losses are:

1. Insulation of the dryer.
2. If the material being dried is a product, there is a tendency to give more attention to the dryer system than if a byproduct is being dried.
3. A large dryer is more important and tends to receive more attention than a smaller system.
4. A small dryer offers per m^3 dryer volume, more m^2 area for convection, and radiation losses.
5. The pressure in the dryer. Convective dryers usually work at a small underpressure (e.g., 5–10 mm wg). An overpressure would lead to uncontrolled emissions and could lead to encrustations at the dryer inlet. However, underpressure leads to the ingress of cooling air.
6. The construction of the dryer. An open design can lead to losses (flash dryer), whereas a closed construction tends to minimize heat losses.

Surprisingly, the steady-state losses are hardly influenced by the temperature of the drying gas. This is because, irrespective of the value of the drying air-inlet temperature, the drying process always occurs at a temperature of 100 ± 50°C.

The steady-state heat losses (normal load) of rotary, flash, and fluid-bed dryers amount to about 20–30% of the enthalpy gain of the feed, whereas in spray dryers the corresponding figure is about 10–15%, with the closed design probably responsible for the difference.

Additional losses

These are caused by starting up the dryer (with the heat used to warm the metal), shut-down of the dryer (subsequent gradual cooling), cleaning of the dryer, and the evaporation of the flushing water and dryer control. Experience has shown that often the desired final-moisture content can be kept constant by maintaining a constant gas-outlet temperature. Many dryers automatically reduce the drying air temperature to maintain a prescribed outlet temperature if the feedflow decreases. The efficiency of the drying process expressed in, e.g., nm^3 of natural gas per metric ton of product decreases. The effect is not very dependent on the air-inlet temperature *per se*. An example follows.

Situation no.	1	2	3	4
TA_{IN} (°C)	160	200	590	800
TA_{OUT} (°C)	80	80	150	150
Efficiency	0.57	0.67	0.77	0.83

The efficiencies are calculated as $(TA_{IN} - TA_{OUT})/(TA_{IN} - 20)$. The example illustrates (a) the decrease of efficiency on lowering TA_{IN} and (b) the independence of the effect on the temperature levels. In the example, the (ΔT = $TA_{IN} - TA_{OUT}$) values were reduced to about two-thirds of the original values in order that the comparison is fair.

The remedy would be to throttle the drying air and to keep TA_{IN} constant. However, this is only possible for spray and rotary dryers since a certain airflow is essential for the operation of flash and of fluid-bed dryers. One might argue that a lower TA_{IN} leads to lower convection and radiation losses as the whole drying process is carried out at a lower temperature. But this is only partly true, as the example shows, and the effect is more than counterbalanced by the efficiency loss.

The additional losses may range from 25–75% of the steady-state total-dryer heat requirement. The lower values are found for fully loaded systems requiring little cleaning, whereas the higher values are found for systems with low throughput, frequent outage and which often require cleaning.

Electrical energy consumption

Three categories of consumption can be distinguished: (a) fan power, (b) motive power, and (c) sundry power.

Fan power

Pressure losses are experienced by the air and evaporated waterflow passing through the dryer system. Pressure losses are mainly located in (a) the air-preparation system, (b) the dryer, and (c) the spent-air cleaning system.

The air-preparation system contains an indirect or direct heater. Since the air flows past finned tubes in an indirect heater, the pressure loss is almost negligible; however, this is not the situation in direct-heater systems.

The pressure loss of the air on passing through a direct heater (usually gas fired) is maximum 50 mm wg when the gas, admitted via a sparger, burns in the total airflow. This type of burner can be used to provide temperatures of up to 500°C. If higher temperatures are required, a combustion chamber is used in which the gas or oil is burned with a 10–20% excess of air compared to the stoichiometry of the reaction. The Δp for this airflow is between 250 and 500 mm wg. Large combustion chambers (e.g., 5,000 kW) exhibit high pressure loss and smaller chambers show lower pressure drop. Secondary air is admitted with the flue gas to obtain the desired drying air temperature and the pressure drop for this flow is about 150 mm wg.

The pressure drop across a dryer is strongly dependent upon the dryer type:

1. It is negligible for a rotary dryer which is, in effect, an empty tube with curtains of falling solids.
2. It is negligible for a flash dryer in which the very diluted air-solids mixture flows through an empty tube.
3. It is substantial for a fluid bed dryer in which the air must overcome the resistance of the distribution plate and the product layer. The resistances of the distribution plate and product layer are approximately equal for a small fluid-bed dryer product layer (0.1–0.2 m), this being necessary to obtain stable fluidization, i.e., no preferential air passages. Deeper beds use a distribution plate offering less resistance than the product layer (Geldart and Baeyens, 1985).
4. It is negligible for a spray dryer which is, in fact, an empty vessel with a very diluted solids-air mixture flowing through it.

The typical spent-air cleaning system may contain one or more of these appliances: a cyclone, a filter, and a scrubber. A cyclone collects particles larger than 5–10 μm, and the associated pressure drop is 100–200 mm wg. A dust filter collects finer particles (to 0.01 μm); the equipment is larger than that of a cyclone and needs normal air velocities with respect to the filtering area of between 1 and 3 m/min. The pressure drop can approach 200 mm wg. A wet scrubber can be very simple, as in the case of a scrubber for air containing sodium chloride, with a pressure drop that is, for all practical purposes, negligible. On the other hand, wet scrubbers have pressure drops of up to 1,000 mm wg.

The spent-air cleaning system processes all solids for flash and spray dryers but only the dust for rotary and fluid-bed dryers. The fans for rotary, flash, and spray dryers can be located in the air-preparation system, in the air-cleaning system, or in both. With fluid-bed dryers it is customary to have fans for air admission and for exhaust.

$$Fanpow = \frac{\phi_v \cdot \Delta p}{3,600 \cdot \eta \cdot 1,000} \text{ kW}$$

where $\eta = 0.6 \times 0.85 \times 0.95 = 0.4845$ (usually taken as 0.5), 0.6 is the fan efficiency, 0.85 is the motor efficiency, and 0.95 takes further losses into account (e.g., lines). Δp should be expressed in N/m^2 (1 mm wg = 10 N/m^2).

Motive power

For rotary dryers rotating at normal speed the motive power is: $0.3 \times \pi/4 \times D^2 \times L$ kW. (Peripheral speed 0.25–0.5 m/sec; see Perry et al., 1984, and for further details, see Chapter 6.)

In flash dryers, a sling or cage mill is required to destroy agglomerates, typically drawing 10–20 kW.

Spray dryers require substantial motive power for rotary wheels or nozzles, e.g., 8 kWhr/t of feed for rotary atomization.

Fluid-bed dryers require a small motive power for vibration if fitted, e.g., 2, 5, or 10 kW.

Sundry power

Electrical power is required for screws, locks, vibrating conveyors, oil pumps, etc. Locks normally require 0.5 kW, whereas vibrating conveyors and screws require between 1 and 10 kW.

Investment

In comparing fixed and variable costs, the concepts of capital charge and maintenance rate should be used.

Example

The investment for a large industrial dryer is $500,000 (U.S.)

$Cap_{CHAR} = 30\%$
$Maint = 5\%$
$Prod = 120,000$ t per annum

Fixed cost per t is:

$$\frac{500,000 \times 0.35}{120,000} = \$1.46 \text{ (U.S.)}$$

This cost can be compared with the energy cost in optimization calculations.

To arrive at an investment figure, an installation cost must be known as well as the capital cost of the equipment. As a rule of thumb, the equipment cost should be multiplied by a factor of 2.75 if the equipment is constructed from carbon steel or by a factor of 2.25 if the equipment is manufactured from stainless steel to obtain the total installed cost (including instrumentation, auxiliaries, allocated building space, engineering, etc.). These figures apply equally to spray, flash, rotary and fluid-bed dryers. In specific cases however, different factors should be used, e.g., an outdoor plant is cheaper than an indoor installation.

Utility prices

In the United States, utility prices (as of 1985) were:

Natural gas

$Cal_{VAL} = 32$ mJ/nm^3
$0.1 per nm^3

Oil

$Cal_{VAL} = 42,000$ mJ per t
$260 per t

Steam

$Cal_{VAL} = 2,000$ mJ per t
$12 per t

Electricity

$0.06 per kWhr

Materials of construction

Carbon and stainless steel are the usual materials of construction for continuous convective dryers.

Material balance (kg/hr)

$$W_{OUT} = 0.01 \times Cap \times A_2$$
$$Sol = Cap - W_{OUT}$$
$$W_{IN} = \frac{Sol \times A_1}{100 - A_1}$$
$$Evap = W_{IN} - W_{OUT}$$

Heat balance (kg/hr)

$$TA_{OUT} = a \cdot TA_{IN} + b$$

For spray drying, $TA_{OUT} = 88.39 \times \log_{10} TA_{IN} + 112.35$ (see Chapter 8). The values of a and b depend upon the type of dryer.

$$Q_1 = Evap(2{,}504 + 1.886 \cdot TA_{OUT} - 4.19 \cdot Tf)$$
$$Q_2 = Sol \cdot CP(TA_{OUT} - c - Tf)$$
$$Q_3 = W_{OUT} \cdot 4.19 \, (TA_{OUT} - c - Tf)$$

The value of c depends upon the dryer type. For more information on CP see Appendix 4.2.

$Qtot_1$ is the total enthalpy change of the components of the process stream.

$$Qtot_1 = Q_1 + Q_2 + Q_3$$

$Qtot_2$ is the total amount of heat which is put into the drying airflow.

$$Qtot_2 = Qtot_1 \times \frac{(TA_{IN} - 10)}{(TA_{IN} - TA_{OUT})} \times d$$

The air-intake temperature is assumed to be 10°C. The value of d depends upon the dryer type and allows for the steady-state heat losses.

Gas flows/fan power

Drying air-preparation system

The presence of a gas burner is assumed and the presence of water in the air is taken to be negligible.

$$Airflo_1 = \frac{Qtot_2}{1.05 \times 1.24\,(TA_{IN} - 10)} \text{ m}^3/\text{hr}$$

Where 1.05 is the specific heat of air in kJ/kg. K (mean 0–600°C).

$$Pow_{G1} = \frac{Airflo_1 \cdot \Delta p}{3{,}600 \cdot 1{,}000 \cdot 0.5} = \frac{Airflo_1 \cdot \Delta p}{1.8 \cdot 10^6} \text{ kW}$$

Spent air-cleaning system

$$Airflo_2 = Airflo_1 \cdot \frac{1.24}{R_A} \cdot f \text{ m}^3/\text{hr}$$

$$R_A = \frac{355}{273 + TA_{OUT}} \text{ kg/m}^3$$

The value of f depends on the dryer type and accounts for the attraction of ingress air. See Chapter 5, Section 5.6 for a discussion of the accuracy of the calculation of $Airflo_2$.

$$WA_{FLO} = \frac{Evap}{R_W} \text{ m}^3/\text{hr}$$

$$R_W = \frac{220}{TA_{OUT} + 273} \text{ kg/m}^3$$

$$Gasflo_2 = Airflo_2 + WA_{FLO} \text{ m}^3/\text{hr}$$

$$Pow_{G_2} = \frac{Gasflo_2 \cdot \Delta p}{1.8 \cdot 10^6} \text{ kW}$$

4.3   Additional remarks

The computer programs concern two cost aspects of the drying step: (a) *Medcos*, the medium cost, in $ per t of product. *Medcos* is directly calculated per t of product. (b) *Powcos*, the power cost, in $ per t of product. *Powcos* is directly calculated per t of product. Hence, the programs do not contain elements that are the same for the four dryer types, e.g., direct labor, indirect labor, instrument air, etc.

Notation

A_1	Feed moisture content	% by wt
A_2	Product moisture content	% by wt
$Airflo_1$	Airflow taken in by the dryer (10°C)	m^3/hr

$Airflo_2$	Airflow leaving the dryer (TA_{OUT})	m^3/hr
a	Constant in $TA_{OUT} = a \times TA_{IN} + b$	
b	Constant in $TA_{OUT} = a \times TA_{IN} + b$	K
Cap	Product mass flow	kg/hr
Cal_{VAL}	Fuel calorific value	mJ/nm^3 or mJ/t
Cap_{CHAR}	Capital charge for investments	%
CP	Solid specific heat	kJ/kg · K
c	Constant in expressions for Q_2 and Q_3	K
D	Rotary dryer diameter	m
d	Constant in expression for $Qtot_2$	
$Evap$	Dryer water evaporation load	kg/hr
$Fanpow$	Fan power	kW
f	Constant in expression for $Airflo_2$	
$Gasflo_2$	Gas flow leaving the dryer	m^3/hr
L	Rotary dryer length	m
$Maint$	Maintenance	%
$Medcos$	Medium cost per metric t of product	$/t
$Powcos$	Power cost per metric t of product	$/t
Pow_{G_1}	Fan power required for $Airflo_1$	kW
Pow_{G_2}	Fan power required for $Gasflo_2$	kW
$Prod$	Annual production	tpa
Q_1	Heat flow for the evaporated water	kJ/hr
Q_2	Heat flow for the dry solids	kJ/hr
Q_3	Heat flow for the residual water	kJ/hr
$Qtot_1$	Net heat	kJ/hr
$Qtot_2$	Steady-state dryer heat requirement	kJ/hr
R_A	Air specific mass	kg/m^3
R_W	Water vapor specific mass	kg/m^3
Sol	Dry solids flow	kg/hr
TA_{IN}	Drying air temperature	°C
TA_{OUT}	Spent drying-gas temperature	°C
Tf	Feed temperature	°C
WA_{FLO}	Evaporated water flow	m^3/hr
W_{IN}	Water flow to the dryer	kg/hr
W_{OUT}	Water flow leaving the dryer	kg/hr
Δp	Pressure loss	N/m^3
$\Delta\rho$	Specific mass difference	kg/m^3
Δ_T	Temperature difference	K
η	Efficiency	
ϕ_v	Gas flow	m^3/hr
ρ_P	Particle specific mass	kg/m^3
ρ_F	Gas specific mass	kg/m^3

Appendix 4.1

```
LIST
10   LPRINT "SETTLING VELOCITY SPHERICAL PARTICLES IN AIR"
20   LPRINT
30   INPUT "TEMPERATURE,DEGREES C";T
40   INPUT "PARTICLE SIZE,M";D
50   INPUT "PARTICLE SPECIFIC MASS,KG/M3";RS
60   LPRINT
70   REM AIR SPECIFIC MASS RA
80   RA=355/(T+273)
90   REM AIR VISCOSITY V (0-100 DEGREES C)
100  V=1.725E-05+4.77E-08*T-2.22E-11*T^2
110  REM DISCRIMINATING PARAMETER CALCULATION,PAR
120  PAR=2*D^3*RA*(RS-RA)*9.810001/(3*V^2)
130  IF PAR>10 GOTO 160
140  U=(RS-RA)*9.810001*D^2/(18*V)
150  GOTO 310
160  IF PAR>200 GOTO 200
170  REM REYNOLDS NUMBER CALCULATION,RE
180  RE=PAR^.8659/10^.9924
190  GOTO 290
200  IF PAR>5000 GOTO 230
210  RE=PAR^.7154/10^.646
220  GOTO 290
230  IF PAR>200000! GOTO 270
240  RE=PAR^.6242/10^.3089
250  GOTO 290
260  REM U IS THE SETTLING VELOCITY IN M/S
270  U=1.76*SQR((RS-RA)*9.810001*D/RA)
280  GOTO 310
290  U=RE*V/(RA*D)
300  GOTO 320
310  RE=RA*U*D/V
320  REM OUTPUT
330  LPRINT
340  LPRINT
350  LPRINT
360  LPRINT USING "TEMPERATURE,DEGREES C                    ###";T
370  LPRINT
380  LPRINT USING "PARTICLE SIZE,M                    #.######";D
390  LPRINT
400  LPRINT USING "REYNOLDS NUMBER,RE                 ########.###";RE
410  LPRINT
420  LPRINT USING "SETTLING VELOCITY,M/S              ##.####";U
430  LPRINT
440  END
0


RUN
SETTLING VELOCITY SPHERICAL PARTICLES IN AIR

TEMPERATURE,DEGREES C? 25
PARTICLE SIZE,M? .00025
PARTICLE SPECIFIC MASS,KG/M3? 2160

TEMPERATURE,DEGREES C              25

PARTICLE SIZE,M                    0.000250

REYNOLDS NUMBER,RE                 26.335

SETTLING VELOCITY,M/S              1.6296


0
```

```
RUN
SETTLING VELOCITY SPHERICAL PARTICLES IN AIR

TEMPERATURE,DEGREES C? 25
PARTICLE SIZE,M? .00075
PARTICLE SPECIFIC MASS,KG/M3? 2160

TEMPERATURE,DEGREES C              25

PARTICLE SIZE,M                    0.000750

REYNOLDS NUMBER,RE               244.144

SETTLING VELOCITY,M/S              5.0358

0

RUN
SETTLING VELOCITY SPHERICAL PARTICLES IN AIR

TEMPERATURE,DEGREES C? 75
PARTICLE SIZE,M? .00025
PARTICLE SPECIFIC MASS,KG/M3? 2160

TEMPERATURE,DEGREES C              75

PARTICLE SIZE,M                    0.000250

REYNOLDS NUMBER,RE                19.956

SETTLING VELOCITY,M/S              1.6199

0

RUN
SETTLING VELOCITY SPHERICAL PARTICLES IN AIR

TEMPERATURE,DEGREES C? 75
PARTICLE SIZE,M? .00075
PARTICLE SPECIFIC MASS,KG/M3? 2160

TEMPERATURE,DEGREES C              75

PARTICLE SIZE,M                    0.000750

REYNOLDS NUMBER,RE               191.661

SETTLING VELOCITY,M/S              5.1862

0
```

Appendix   4.2

Specific heat of solids

Most textbooks pay little attention to the specific heat of solids. Values can be found in the literature or alternatively Kopp's Law, which is a method for quick approximation, can be used.

Kopp's Law states that the specific heat of a solid at room temperature is approximately equal to the sum of the atomic heat capacities divided by the molecular weight.

kJ/kilo-atom · K

Carbon	7.5	Oxygen	16.8
Hydrogen	9.6	Phosphorus	22.6
Boron	11.3	Fluorine	21.0
Silicon	15.9	Other elements	26.0

(From Raymond Division, Combustion Engineering, Inc., Chicago, Illinois)

For example to calculate the specific heat of sodium sulphate (Na_2SO_4)

Element	Atomic weight	Atomic heat capacity
Na_2	46.00	52.0
S	32.06	26.0
O_4	64.00	67.2
Total	142.06	145.2

Specific heat = 145.2/142.06 = 1.022 kJ/kg·K.

Perry's *Chemical Engineers' Handbook* (6th ed.) gives 0.986 kJ/kg · K, that is, a 5% error from the value derived from Kopp's Law. The law is also applicable to hydrates.

References

Coulson, J. M. and J. F. Richardson (1968). *Chemical Engineering Volume II,* Pergamon, London.

Geldart, D. and J. Baeyens (1985). The design of distributors for gas-fluidized beds, *Powder Technol.,* 42, 67.

Nonhebel, G. and A. A. H. Moss (1971). *Drying of Solids in the Chemical Industry,* Butterworths, London.

Peters, M. S. and K. D. Timmerhaus, (1968). *Plant Design and Economics for Chemical Engineers,* McGraw-Hill, New York

Perry, R. H., D. W. Green, and J. O. Maloney (1984). *Perry's Chemical Engineers' Handbook,* McGraw-Hill, New York.

Swenson Inc. (1985). Private Communication.

Williams-Gardner, A. (1971). *Industrial Drying,* Leonard Hill, London.

5

CONTINUOUS FLUID-BED DRYING

5.1 Introduction

Section 5.2 gives a general introduction to the subject. In the section that follows, theoretical aspects of fluidity are covered, e.g., the concept of minimum fluidization velocity is discussed. Theoretical aspects of drying are dealt with in Section 5.4, along with a brief review of Mollier's H-y diagram for air and water. Drying theory is used to size a rectangular dryer with a shallow bed for a drying operation in which the bulk of the moisture is free. A second example considers the situation of bound water. Section 5.5 discusses the design of a circular fluid-bed dryer with an example to illustrate the procedure. Section 5.6 explains the computer program in Appendix 5.1 for the design of a rectangular fluid-bed dryer where the design method differs slightly from the first example given in Section 5.4. Lastly, investment considerations are covered in Section 5.7.

5.2 General description

A continuous fluid-bed dryer is essentially equipment in which a continuous feed of wet particulate material is dried by contact with warm or hot air that is blown through to maintain the material in a fluidized state. Because the drying action depends upon fluidization, the moisture content of the feed is normally lower than that used in flash and in rotary dryers.

Two principal types fluid-bed dryers are in use: a circular type with a deep bed (0.5–2.0 m) and a rectangular form that usually has a bed up to 0.2 m deep.

Circular fluid-bed dryers were introduced in the United States and in Canada in 1948. Operating with a high air-inlet temperature (e.g., 800°C), they were used to dry coal, limestone, blast furnace slag, and similar materials (Beeken, 1960). The equipment is not vibrated and the fluidized bed has the characteristics of a continuous stirred tank reactor. The composition of the product stream is the same as the composition of the bed. A consequence is that incrustation is not likely to occur since the dryer contents are dry. Neither are there any restrictions in scaling up the apparatus. Although the electrical consumption is high, since the fan drive is used to fluidize the bed, yet one of the principal attractions of this system is its simplicity (see Fig. 5.1).

The mode of operation of the rectangular fluid-bed dryer is shown in Figure 5.2. Usually the dryer has a small bed height leading to a plugflow characteristic. It is not easy to accomplish fluidization of the wet material near the feed inlet, and this can lead to improper transportation and incrustation. Various modifications can be made to remedy this fault: (a) more air in the feed section, (b) directing the blown air so as to create a transport effect, and (c) vibration of the dryer. Frequently, the latter option is found because this equipment was developed from vibrating conveyors that can be used for many types of particulate materials and, hence, is applicable to a wide range of particle sizes. Vibration safeguards the transport independently of the air supply and prevents the occurrence of incrustations and may lead to superior heat and mass transfer. However, it limits scaling up to a maximum size of, e.g., 8×1.6 m^2. The maximum temperature for vibrating equipment is approximately 300°C. Rectangular dryers with deep beds (e.g., 0.5–1 m) exhibit considerable backmixing and therefore relatively dry material can be found in the feed section, which decreases the tendency for incrustation and makes vibration superfluous (Tailor, 1969).

To some extent, fluid-bed dryers are intermediate between direct-heat rotary and flash dryers:

1. Flash dryers are sensitive to load variations whereas direct-heat rotary dryers are relatively insensitive.
2. Flash dryers have residence times of a few seconds, direct-heat rotary dryers approximately thirty minutes, while small bed-height fluid-bed dryers have a residence time of several minutes.
3. The moderate feed disintegration is intermediate to that caused in flash dryers (maximum feed disintegration) and rotary dryers (mild disintegration).

The size of fluid-bed dryers becomes disproportionally large for materials having a weight-average particle size of ≤ 100 μm. Although flash dryers do not suffer from this restriction, rotary dryers are restricted to some

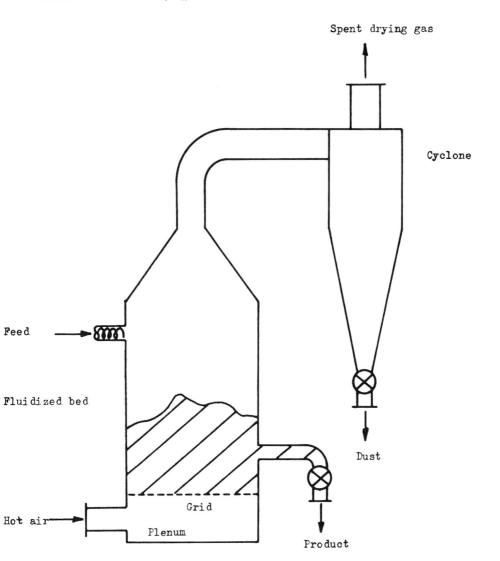

Figure 5.1   Round fluid-bed dryer.

extent; which is another reason for the feed moisture content being lower
in fluid-bed dryers than in flash or in rotary since less water is attached to
larger particles.

   An interesting aspect is the installation of heat-exchanging surfaces in
tall fluid beds; the heat transfer is good and the heat transferred by conduc-

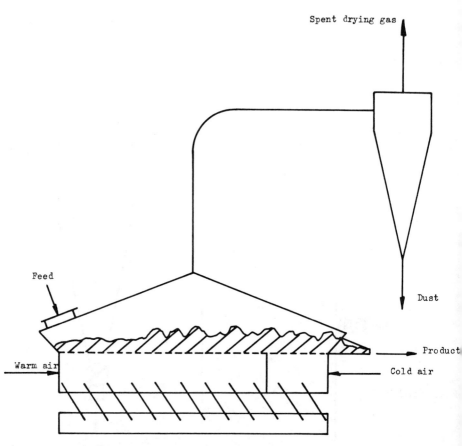

Figure 5.2 Rectangular fluid-bed dryer (vibrated).

tion decreases the amount of required heated air. In principle, heat transfer via conduction is more effective than via convection but the heat-exchange surfaces must not suffer from incrustation. A typical application is the use of rectangular fluid beds with heating panels in the polymer industry. Typical values for the wall-to-bed heat-transfer coefficients are in the range 200–600 W/m² · K. It is recommended that actual heat-transfer coefficients are established by means of drying tests rather than reliance on published correlations. It is possible to combine a fluid-bed dryer and a fluid-bed cooler in the same equipment which cannot be done with a flash dryer, and which requires special internals in a rotary dryer. Also fluid-bed dryers can combine drying and particle-size enlargement, i.e., granulation.

Fluid-bed granulation can be defined as a particle-forming process that is achieved by spraying a liquid into a fluidized bed already containing particles. The liquid may contain dissolved material.

There are two possible nucleation mechanisms: (a) the sprayed droplets may evaporate and dissolved matter may be left behind or (b) attrition may occur between fluidized granules.

There are also two possible growth mechanisms: (a) a wetted particle may attach itself to a second particle by liquid bridging which with subsequent drying creates a solid bridge (*agglomeration*) or (b) the wetted particle may dry and the residual matter may attach to the original particle (*layering*). Fluid-bed granulation is capable of producing particulate solids with an average size of between 0.5 and 3 mm. However, adding all the solid material to be crystallized in a solution is an energy-intensive process, because the thermal efficiency of a convective dryer is less than the thermal efficiency of a one-stage evaporator. A four-stage evaporation unit can evaporate approximately 3 kg of water with 1 kg of steam. A further option for fluid-bed processing is coating.

5.3   Fluidization theory

The term *fluidization* was coined to describe a certain mode of contacting granular solids with fluids. To illustrate, let the solid be a well-rounded silica sand contained in a cylindrical vessel with a porous bottom. The fluid is air. As the air passes upward, through the porous bottom and the bed, measurements can be taken of the bed height and of the pressure drop across the bed as a function of the airflow rate. A profile as shown in Figure 5.3 will be obtained.

When the pressure drop reaches its maximum value, the bed starts to expand. In this condition, the individual sand granules are disengaged from each other and may be moved readily around by the expenditure of much less energy than would be required if the bed were not suspended by the airstream. The mobility of the aerated sand column resembles that of a high-viscosity liquid. This condition is known as the *fluidizing point* and the corresponding airflow rate is termed the *minimum fluidization velocity*. With a further increase of the airflow, the point at which the first bubbles appear can be observed. This point locates the minimum bubble point and the minimum bubble-point velocity. A further increase in the flow rate leads to the maximum bed-expansion ratio.

The first large-scale fluidization application was used in the United States in 1940 for cracking oil vapor. Fluid cracking catalysts have a weight average particle size of about 80 μm and a bulk density of about 1,100 kg/m^3.

The description and successful operation of the fluid cracking units precipitated a number of fundamental and applied studies in fluidization

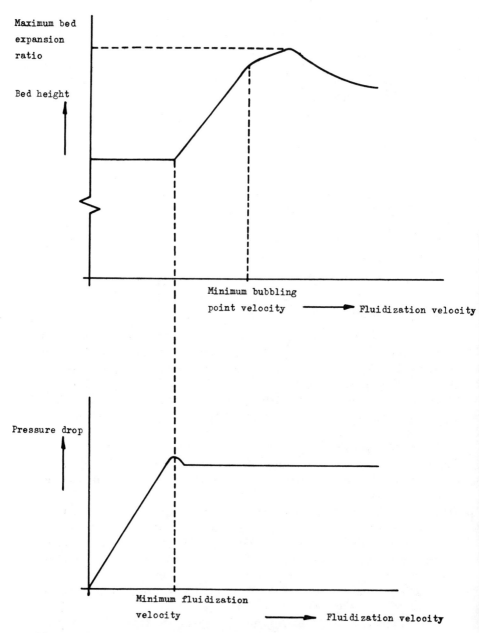

Figure 5.3   Diagrams for ideally fluidizing solids.

fields. The very efficient heat and mass transfer in a fluidized bed was seen as the principal advantage of this system. Figure 5.3 can be used to show how granular materials display ideal fluidization characteristics; however, not all materials exhibit this behavior. For example, channeling (preferential airflow) is a frequently found unwanted phenomenon, and Figure 5.4 shows typical pressure drop/flow data for moderately channeling solids (with flour, particle size ≤ 20 μm, exhibiting the channeling). The design of the gas-inlet device has a profound effect on channeling; thus, with porous plates, by means of which gas distribution into a bed tends toward uniformity, channeling tendencies are much reduced when compared with a multi-orifice distributor in which the gas is introduced through a relatively small number of geometrically spaced holes.

Stable shallow fluidized beds require a pressure drop across the gas-distributing device of between one-third and one-half of the pressure drop across the bed.

Example 1

With a bed height of 0.1 m, containing a material of bulk density of 1,000 kg/m³, the pressure drop across the bed is equal to $1,000 \times g \times 0.1 \approx 1,000$ N/m² (100 mm wg).

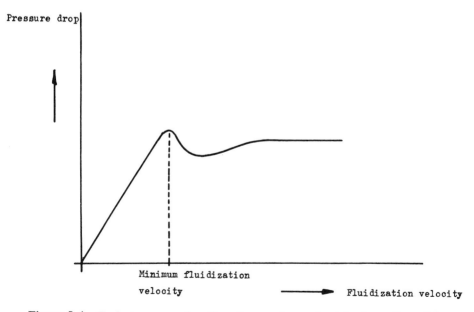

Figure 5.4 Typical pressure drop/flow diagram for moderately channeling solids.

The pressure drop across the distributing device must be about 400 N/m²
to avoid preferential gas passage through the bed.

Higher beds, e.g., between 0.5 and 1 m require lower pressure drops
across the bottom plate. Here it can be between 10 and 20% of the bed
pressure drop.

An important factor known to affect channeling is the moisture content
of the solid. Two types of fluid-bed dryers can be distinguished: (a) rectangular types with plugflow and usually a shallow product layer and (b)
cylindrical vessels with bed heights of 0.5–2 m operated as continuous
stirred tank reactors.

A shallow, wet bed of particulate material is hard to fluidize. As the
average moisture content decreases, fluidization becomes viable. However, particle sizes of 1 mm or larger exhibit spouting beds rather than fluid
beds. It is not customary to distinguish between these phenomena; an
upward flow through a particulate mass to be dried means we are dealing
with a fluid-bed dryer if the material moves. The reason undoubtedly is
that the terms *fluid bed* and *fluidization* are generally associated with efficient mass and heat transfer.

The ability to predict reliably the point of incipient fluidization is of
basic importance in virtually all fluidized-process studies and designs.

Fluidized reactors and dryers are usually operated at fluid rates that are
well in excess of minimum fluidization rates. However, the fluid rate should
not exceed the terminal velocity leading to the dilute phase. Leva (1959)
recommended for the estimation of the minimum fluidization mass velocity:

$$G_{mf} = 688.\overline{d}_p^{1.82} \frac{[\rho_F(\rho_S - \rho_F)]^{0.94}}{\mu^{0.88}} \quad (\mu \text{ in } cP)$$

In SI units:

$$G_{mf} = 0.0093.\overline{d}_p^{1.82} \frac{[\rho_F(\rho_S - \rho_F)]^{0.94}}{\mu^{0.88}} \text{ kg/m}^2 \cdot \text{sec}$$

This equation has been supported by many experimental studies. The correlation is limited to applications where:

$$Re = \frac{G_{mf} \cdot \overline{d}_p}{\mu} < 10.$$

If $Re > 10$, G_{mf} must be corrected using Figure 5.5. G_{mf} must be multiplied
by the value found. If the material is vesicular, the solid specific mass must
be corrected. The composite diameter must be calculated according to:

$$\overline{d}_p = \frac{1}{\Sigma \frac{X}{d_p}}$$

Correction factor

for G_{mf}

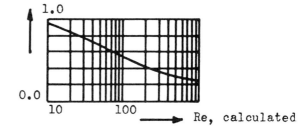

Re, calculated

Figure 5.5 Correction factor for results obtained with Leva's equation if $Re > 10$. (Courtesy of Génie Chimique.)

Example 2

A perforated plate is charged with dry vacuum-pan salt (NaCl). The bed height is 0.2 m and the bulk density is 1,200 kg/m³. The specific mass is 2,160 kg/m³. Sieve analysis provides the following results:

Microns	% by wt	X	d_p	X/d_p
>600	0	0	—	—
500–600	9	0.09	550	0.000164
420–500	31	0.31	460	0.000674
250–420	35	0.35	335	0.001045
125–250	20	0.20	187.5	0.001067
63–125	5	0.05	94	0.000532
<63	0	0	—	
				————+
				0.003482

$$\bar{d}_p = \frac{1}{0.003482} = 287 \ \mu m$$

The salt is fluidized with air at 20°C, and the air discharges from the bed under atmospheric pressure. Fluidization pressure drop $\Delta p = \rho_b \cdot g \cdot h = 1,200 \cdot 9.81 \cdot 0.2 = 2,354 \ N/m^2$.

Inlet-air pressure: $100,000 + 2,354 = 102,354 \ N/m^2$

$$\rho_F = 1.3 \times \frac{273}{273+20} \times \frac{102,354}{100,000} = 1.24 \ kg/m^3$$

$$G_{mf} = 0.0093 \cdot (2.87 \cdot 10^{-4})^{1.82} \frac{[1.24(2,160 - 1.24)]^{0.94}}{(18 \cdot 10^{-6})^{0.88}} = 0.083 \ kg/m^2 \cdot sec.$$

$$\text{Velocity } v_{mf} = \frac{0.083}{1.24} = 0.067 \ m/sec$$

$$Re = \frac{G_{mf} \cdot \bar{d}_p}{\mu} = \frac{0.083 \cdot 2.87 \cdot 10^{-4}}{18 \cdot 10^{-6}} = 1.32$$

No correction required.

A more recent correlation concerning the minimum fluidization velocity (Baeyens and Geldart, 1973) is:

$$v_{mf} = \frac{0.0009(\rho_S - \rho_F)^{0.934} g^{0.934} \cdot \bar{d}_{sv}^{1.8}}{\mu^{0.87} \cdot \rho_F^{0.066}} \text{ m/sec}$$

However, for the purpose of this text, it is not necessary to investigate the concept of minimum fluidization velocity in any further detail. Similar empirical relationships exist for the minimum bubbling-point velocity. The fluidization velocity actually chosen is usually many times greater than the minimum fluidization velocity. Table 5.1 (Perry et al., 1984) provides an estimate of the maximum superficial air velocities obtained through vibrating conveyor screens.

Using this table, we find that the maximum superficial air velocity for the material of the example is about 20 times that of the minimum fluidization velocity. For relatively small particles, this is not unusual, but the factor is smaller for larger particles.

The fluidization velocity is often chosen after consideration of the carry-over of fines by calculating the terminal velocity of the fines. However, the elimination of fines is time-dependent.

Table 5.1 Estimate of Maximum Superficial Air Velocities Through Vibrating Conveyor Screens (m/sec)

Mesh size	Microns	Specific gravity	
		2.0	1.0
200	74	0.22	0.13
100	149	0.69	0.38
50	297	1.4	0.89
30	590	2.6	1.8
20	840	3.2	2.5
10	2,000	6.9	4.6
5	4,000	11.4	7.9

Source: Columns 1, 3, and 4: Courtesy Carrier Vibrating Equipment, Inc. Column 2 gives the corresponding hole sizes in microns. The mesh size reflects the U.S. number, National Bureau of Standards LC-584 and ASTM E-11.

5.4 Drying theory

The drying process in a rectangular-type dryer with plugflow and a shallow product layer will now be considered. Figure 5.2 depicts a typical drying plant, with the wet product entering at the left and moving to the right; drying, postdrying and cooling occur subsequently. Surface moisture (free moisture) is evaporated on the extreme left of the bed. Both the gas flow and the product attain the adiabatic saturation temperature, i.e., the temperature the gas acquires on cooling down and accepting the water vapor. The gas sensible heat is converted into the latent heat of evaporation.

This part of the bed ends when the product has reached the critical moisture content. The next section of the dryer is used to bring down the bound moisture to the desired moisture content, as along the section the material temperature increases.

Cooling occurs along the right-hand section of the bed. The plugflow dryer offers the option of simply cooling the process flow in the same piece of equipment, something that cannot be achieved in either a flash or spray dryer but can be realized in a rotary dryer. Often cooling is required to prevent caking of the product in the bunkers or silos which is caused by interactions with the atmosphere. Figure 5.6 shows a normal temperature and moisture profile of the dryer. To size the first part (drying), Mollier's *H-y* diagram is used, a small part of which is reproduced in Figure 5.7. The design equation is:

$$\phi_{H_2O} = B \cdot L_1 \cdot v_F \cdot \rho_F (y_0 - y_i) \quad \text{kg/sec} \tag{5.1}$$

y_0 and y_i are found as follows (see Fig. 5.7): atmospheric air enters at 10°C with a relative saturation of 30% followed by heating in an indirect heater to 50°C. The drying process *per se* is represented by the sloped line (isenthalpic change of state). The situation shown in Figure 5.7 is the evaporation of water from a solution of the solid in water, with the solution exerting 90% of the saturated vapor pressure of water.

$y_0 = 14.10^{-3}$ kg of water per kg of dry air
$y_i = 2.10^{-3}$ kg of water per kg of dry air
v_F can be selected from either Table 4.1 or Table 5.1 or can be found using Appendix 4.1 in Chapter 4.

v_F is the velocity of the warm gas. Cooling down to 21°C causes a lower velocity. A further reduction may occur by widening the freeboard.

The design equation leads to the value of L_1, the length of the first part of the bed. An implicit assumption is that the feed enters at the adiabatic saturation temperature. L_1 is multiplied by, e.g., 1.15 to allow for the steady-state heat losses.

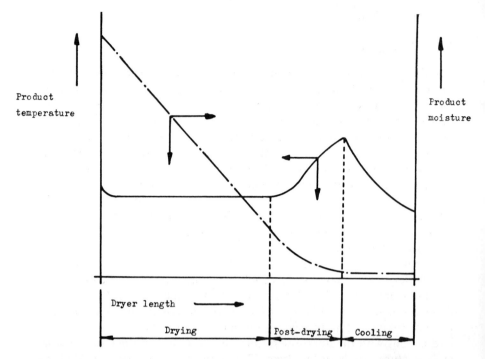

Figure 5.6 Temperature and moisture profile for a continuous rectangular fluid-bed dryer.

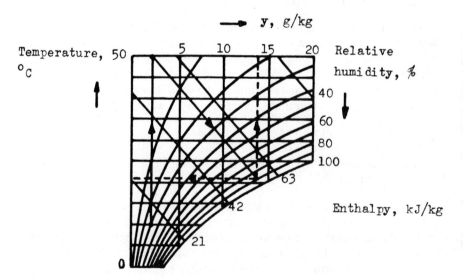

Figure 5.7 A small part of Mollier's *H-y* diagram for air/water.

Table 5.2 **Applicability of the Model for the Evaporation of Free Water on Fluid-Bed Drying**

Product	Scale of test[a]	h_{exp}, mm	v_F, m/sec	TA_{IN}, °C	A_1, % by wt
Cubic vacuum pan salt	p	150	0.9	85	2.0
Dendritic salt	p	100	1	180	15
Open pan salt	i	150[b]	0.4	110	15
Solid peroxy-di-carbonate (batchwise drying)	i	700[b]	0.7	40	20–25

[a]p: pilot-plant scale; i: industrial scale.
[b]Approximate.

Table 5.3 **Physical Properties of Fluid-Bed-Dried Materials**

Product	d_p, mm	ρ_S, kg/m^3	ρ_b, kg/m^3	CP, kJ/kg·K
Cubic vacuum pan salt	0.4	2,160	1,250	0.870
Dendritic salt	0.2	2,160	800	0.870
Granular salt	1.5	2,160	1,250	0.870
Open pan salt	0.7	2,160	800	0.870
Solid peroxy-di-carbonate	0.1	1,100	450	—
Anhydrous citric acid	0.45	1,665	900	1.210
Citric acid monohydrate	0.65	1,540	800	1.470
KAS (A fertilizer)[a]	2	1,389	915	2.0[b]
Sand	—	2,600	—	0.840

[a]Data from Winterstein et al., 1964
[b]Approximate.

Table 5.2 summarizes cases where the simple model was found to be applicable. Vaněček et al. (1966 p. 118) stated that the adiabatic saturation temperature is usually obtained. Table 5.3 summarizes the physical properties of materials mentioned in Tables 5.2, 5.4, and 5.5.

Sizing of the second part (postdrying)

In order to simplify the situation, assume that heating occurs only in crossflow and that no additional moisture is evaporated. Figure 5.8 illustrates the starting point of an ideally mixed crossflow.

A heat balance for a small length *dx:*

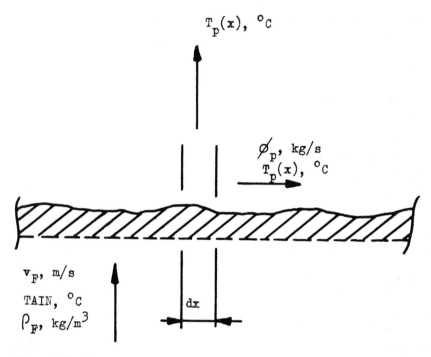

Figure 5.8   Crossflow air heating of particulate material.

$$\rho_F \cdot v_F \cdot B \cdot dx \cdot C_p \, [TA_{IN} - T_p(x)] = \phi_p \cdot CP \cdot dT_p(x)$$

Integrating between $x = 0$ and $x = L_2$ yields:

$$L_2 = \frac{CP \cdot \phi_p}{C_p \cdot v_f \cdot \rho_F \cdot B} \, \log_e \left[\frac{T_p(x=0) - TA_{IN}}{T_p(x=L_2) - TA_{IN}} \right] \text{ m} \qquad (5.2)$$

L_2 is multiplied by 1.15 to allow for the steady-state heat losses. Table 5.4 relates to the applicability of the combination of the model for the evaporation of free water and the model for the evaporation of bound moisture.

   Table 5.5 contains data concerning the applicability of a crossflow cooling model; using equation (5.2) and multiplying L_3 by 1.15. Figure 5.9 depicts a typical vibrating fluid-bed-dryer system.

Example

We have to obtain 20 t/hr of dried vacuum-pan salt from a continuous fluid-bed dryer. The feed enters with a water content of 2.5% by weight at a temperature of 50°C. The product must contain less than 0.1% by weight.

   Experience has shown the need to heat the salt to 65°C to attain the required low moisture content. The weight-average particle size is 400 μm.

Table 5.4 Applicability of the Combination of the Models for the Evaporation of Free Water and the Heating-up in Crossflow on Fluid-Bed Drying

Product	Scale of test[a]	h_{exp}, mm	v_F, m/sec	TA_{IN}, °C	A_1, % by wt
Cubic vacuum-pan salt	i	60–200	1.0–1.1	165–210	2.0
Cubic vacuum-pan salt	i	150	0.85	110–170	1.9
Cubic vacuum-pan salt	i	200	1.25	160	2.4
Cubic vacuum-pan salt	p	150	0.8	95	0.9
Cubic vacuum-pan salt	p	150	1.4	62	2.7
Cubic vacuum-pan salt	i	150	0.4	110	3.5
Dendritic salt	p	100	1	165	19
Granular salt	p	150	1.7	108	1.25
Anhydrous citric acid	p	100	1/1.2	94/60	2.0
Citric acid monohydrate	p	70/200	1.25/0.95	60	2.5

[a]p: pilot-plant scale; i: industrial scale.

Table 5.5 Applicability of the Model for Crossflow Cooling in a Plugflow Fluid-Bed Dryer

Product	Scale of tests[a]	h_{exp}, mm	v_F, m/sec	TA_{IN}, °C
Anhydrous citric acid	p	100	0.7/1	15
Citric acid monohydrate	p	70/200	1	15
KAS (a fertilizer)[b]	i	[c]	1.2	34
Cubic vacuum pan salt	i	200	0.9	26

[a]p: pilot-plant scale; i: industrial scale.
[b]Data from Winterstein et al., 1964.
[c]Not stated.

Salt has a specific heat of 0.87 kJ/kg · K; ambient air is heated indirectly to 170°C with steam. The specific heat of air is 1.0 kJ/kg · K and the water vapor pressure of saturated brine is 75% of the water vapor pressure between 0 and 100°C.

Product flows (kg/hr)

	In	Out
Water	512	20
Salt	19,980	19,980
	⎯⎯⎯ +	⎯⎯⎯ +
Total	20,492	20,000

Evaporated water = 512 − 20 = 492 kg/hr = 0.1367 kg/sec

Figure 5.9 Vibrating fluid-bed-dryer system in the manufacturer's shop. (Courtesy of Carrier Vibrating Equipment, Louisville, Kentucky.)

Sizing the first part

In Equation (5.1)

$v_F = 1.0$ m/sec (see Table 5.1)

$$\rho_F = 1.29 \times \frac{273}{273+170} = 0.795 \text{ kg/m}^3$$

$B = 1.25$ m

$y_0 = 0.002$ kg/kg

$y_i = 0.051$ kg/kg (Mollier's H-y diagram)

$$L_1 = \frac{0.1367}{1.25 \times 1.0 \times 0.795(0.051 - 0.002)} = 2.81 \text{ m}$$

Taking a length of $1.15 \times 2.81 = 3.23$ m

Sizing the second part

The adiabatic saturation temperature is 46°C.

Equation (5.2)

$$L_2 = \frac{0.870 \times 5.55}{1.0 \times 1.0 \times 0.795 \times 1.25} \log_e \left(\frac{46-170}{65-170} \right) = 0.81 \text{ m}$$

Taking a length of 1.15 × 0.81 = 0.93 m

Note that the sensible heat made available by cooling the feed from 50 to 46°C is neglected.

Sizing the third part

It must be possible to cool the salt to 40°C with air of 25°C.

Equation (5.2)

$$L_3 = \frac{0.870 \times 5.55}{1.0 \times 1.0 \times 1.18 \times 1.25} \log_e \left(\frac{65-25}{65-40} \right) = 1.54 \text{ m}$$

Taking 1.15 × 1.54 = 1.77 m

$$\rho_F = 1.29 \times \frac{273}{273+25} = 1.18 \text{ kg/m}^3$$

The total dryer/cooler length is 3.23+0.93+1.77=5.93 m and the width is 1.25 m.

Steam consumption

Steam consumption is based on an ambient temperature of 10°C. The drying airflow is: (3.23+0.93)1.25 × 1.0 × 0.795 × 3,600 = 14,882 kg/hr. The presence of water vapor in the atmospheric air of 10°C is neglected: 14,882 × 1.0(170 − 10) = 2,381,120 kJ/hr, which is exchanged in the air heater. Heat losses from the air heater are neglected.

The latent heat of evaporation of steam of 14 bar is 1,964.7 kJ/kg.

$$\frac{2,381,120}{1,964.7} = 1,212 \text{ kg of steam/hr}$$

$$\frac{2,381,120}{492} = 4,840 \text{ kJ/kg of evaporated water.}$$

$$\frac{1,212}{492} = 2.46 \text{ kg of steam/kg of evaporated water.}$$

The long-term consumption (including start-up, shut-down, cleaning, and low load) is obtained by multiplying by 1.3.

1.3 × 2.46 = 3.20 kg of steam / kg of evaporated water.

$$3.20 \times \frac{492}{20} = 78.7 \text{ kg of steam / metric t of salt}$$

1.3 × 4,840 = 6,292 kJ/kg of evaporated water.

Electrical energy consumption

Drying airflow   The drying-air fan power consumption is based on an ambient temperature of 25°C.

$\rho_F = 1.18$ kg/m^3

$\dfrac{14,882}{1.18} = 12,612$ m^3/hr (drying airflow)

Bed height = 0.2 m

Salt bulk density = 1,250 kg/m^3

Pressure loss in the bed is $1,250 \times 0.2 \times 9.81 = 2,452.5$ N/m^2

Pressure loss in the distribution plate is $0.5 \times 2,452.5 = 1,226$ N/m^2

$$P = \frac{12,612 \times 3,678.5}{3,600 \times 1,000 \times 0.5} = 25.8 \text{ kW}$$

Installed $1.15 \times 25.8 \approx 30$ kW

Cooling airflow

$1.77 \times 1.25 \times 1.0 \times 1.18 \times 3,600 = 9,399$ kg/hr

$\dfrac{9,399}{1.18} = 7,965$ m^3/hr

$$P = \frac{7,965 \times 3,678.5}{3,600 \times 1,000 \times 0.5} = 16.3 \text{ kW}$$

Installed $1.15 \times 16.3 \approx 20$ kW

Exhaust fan   The mass flow of drying and cooling air is $14,882+9,399$ = 24,281 kg/hr. The total air mass flow that is handled is $1.1 \times 24,281 =$ 26,709 kg/hr (factor 1.1 is due to the attraction of ingress air). The exhausted gases will be at about 55°C.

$$\rho_F(\text{air}) = 1.29 \times \frac{273}{273+55} = 1.074 \text{ kg/m}^3$$

Exhausted air volume $\dfrac{26,709}{1.074} = 24,869$ m^3/hr

Water vapor mass flow 492 kg/hr

$$\rho_F(\text{water vapor}) = 0.8 \times \frac{273}{273+55} = 0.666 \text{ kg/m}^3$$

$\dfrac{492}{0.666} = 739$ m^3/hr

Total gas flow is $739+24,869 = 25,608$ m^3/hr

Cyclone pressure drop is 1,500 N/m^2

$$P = \frac{25,608 \times 1,500}{3,600 \times 1,000 \times 0.5} = 21.3 \text{ kW (with a 25-kW motor installed)}$$

Vibration: 10 kW consumption
Rotary locks: $2 \times 1.5 = 3$ kW consumption

Total electrical-energy consumption is

$$25.8 + 16.3 + 21.3 + 10 + 3 = 76.4 \text{ kW}$$

$$\frac{76.4}{20} = 3.82 \text{ kWhr/metric ton of salt}$$

Long-term consumption figure is $1.3 \times 3.82 = 5.0$ kWhr/t.

Variable costs

One metric ton of steam costs $12.00, whereas one kWhr costs 0.06.

Steam costs: $78.7 \times 10^{-3} \times 12 = 0.94$

Electricity costs: $5.0 \times 0.06 = \underline{0.30} +$
$\$1.24/\text{t of salt}$

Fixed costs

If we refer to Section 5.7, the price for the dryer may be $50,000. We then have to triple this sum to allow for fans, ducts, the heater, and the cyclones: $150,000. Multiply this figure by 2.25 to arrive at the total investment: $337,500. Thus, the fixed costs per metric ton of salt (8,000 hr per annum) are: $(337,500 \times 0.35)/(8,000 \times 20) = \$0.74$.

Example*

We wish to obtain 500 kg of a granular organic formula from a continuously functioning rectangular fluid-bed dryer. The feed enters the dryer with a water content of 25% by weight at 15°C. The product moisture content must be less than 0.5% by weight. 50% by weight of the particles is larger than 1,750 μm, with no particles smaller than 1,000 μm. The maximum allowable product temperature is 100°C. The material's specific heat is 0.95 kJ/kg · K and the material itself is insoluble in water. The "wet" bulk density is 595 kg/m^3, and the "dry" bulk density is 523 kg/m^3. The specific heat of the air is 1.0 kJ/kg · K.

Screening experiments in a small, circular batch fluid-bed dryer are carried out.

---

*The previous example deals with a drying duty in which the bulk of the moisture is free moisture, which is not always the case. It may be that the bulk of the water is bound. This example shows how to proceed under such circumstances.

Fluid-bed diameter: 0.2 m
Feed weight: 1.85 kg

Initial stationary product-height calculation

Feed volume: 1.85/0.595 = 3.1 l
Product height: $3.1/(\pi/4 \times 2^2) = 1$ dm (0.1 m)

An optimum set of conditions is selected. The air velocity below the distribution plate is 2.0 m/sec and the air temperature is 90°C. By maintaining these conditions for 10 minutes, an in-spec product is obtained. In this period of 10 minutes, the following observations are made:

1. The bed's temperature gradually rises from 15°C to 90°C.
2. The product's water content simultaneously diminishes in weight from 25% to 0.2%.
3. Some dust carry-over occurs.
4. Particle breakage does not occur.

The optimum set of conditions is used for scaling up.

Drying
Mass Balance Experiment

	Kg	
	In	Out
H_2O	0.46	0.00
Solid	1.39	1.39
	———+	———+
Total	1.85	1.39

Water evaporation 0.46 kg. Specific water evaporation capacity $0.46 \times 60/(10 \times \pi/4 \times 0.2^2) = 87.9$ kg/hr · m^2

Industrial Duty Mass Balance

	Kg/hr	
	In	Out
H_2O	167	0
Solid	500	500
	———+	———+
Total	667	500

Water evaporation 167 kg/hr
Area required 167/87.9 = 1.90 m^2
Take $1.15 \times 1.90 \approx 2.2$ m^2

Cooling

A subsequent cooling from 90 to 40°C is desired. The maximum ambient temperature is 30°C. Equation (5.2) is referred to. Air velocity is 1.5 m/sec. Air specific mass at 30°C = $1.29 \times 273/(273 + 30) = 1.16$ kg/m^3

$$\text{Area} = \frac{0.95 \times 500}{3,600 \times 1.0 \times 1.5 \times 1.16} \log_e \left(\frac{90-30}{40-30} \right) = 0.14 \text{ m}^2$$

Take 0.2 m^2.

Total dryer and cooler sizes

2.2 + 0.2 = 2.4 m^2. Take 4×0.6 m^2
Dryer length: 3.65 m
Cooler length: 0.35 m

Calculations of steam and of electrical-energy consumption follow the steps of the previous example. The transport and drying functions should be checked in a pilot dryer (size, e.g., 3×0.2 m^2).

5.5 Circular fluid-bed dryers

Section 5.2 contains the introductory remarks and also an illustration depicting such a dryer (Fig. 5.1). The composition of the material in the bed equals the composition of the material in the exit flow. Because the contents are almost ideally mixed, a substantial amount of the material has a residence time of, e.g., 10% of the average residence time. Thirteen sets of data (Williams-Gardner, 1971; Beeken, 1960 and Vreeland and Bacchetti, 1982) were used to establish the relationship

$$TA_{OUT} = 0.0141 \times TA_{IN} + 88.8 \text{ °C}.$$

The method used is linear regression; however, the correlation coefficient is only 0.135 (see Fig. 5.10). The temperature of the fluidized bed is very uniform, thus permitting the performance of subtle drying operations (Vreeland and Bacchetti, 1982; Viebrock and Hodel, 1983). In both of these references, the replacement of a direct-heat rotary dryer by a circular fluid-bed dryer is described. Both undercuring and overcuring are impossible because of the accurate temperature control (large heat capacity).

It is not possible to dry and cool with the same equipment. An important consideration is the substantial power consumption of the fan. Figure 5.11 depicts a circular fluid-bed dryer and also a circular fluid-bed cooler. Note that a flash dryer is being used as a predryer here.

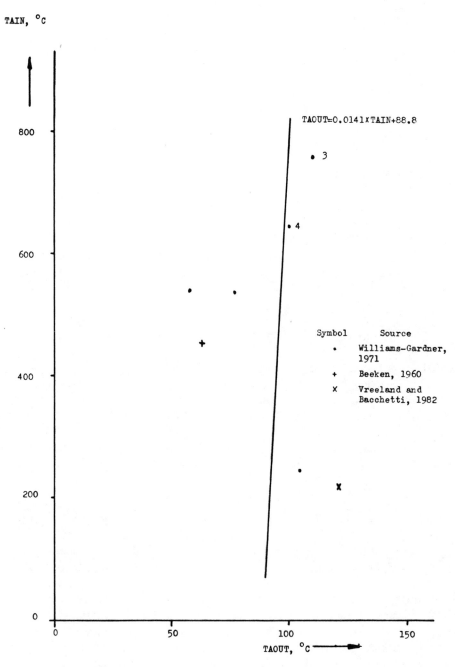

Figure 5.10   Relationship between air-inlet and air-outlet temperatures for circular fluid-bed dryers.

Example

The duty is to obtain 40 metric tons per hr of sand from a continuously functioning circular fluid-bed dryer. The feed entering contains 5% water by weight and is at 10°C. The moisture content of the product must be less than 0.1%. Average particle size is 400 μm. Specific heat of the sand is 0.840 kJ/kg · K.

Ambient air is heated to 750°C by burning natural gas. Specific heat of the air is 1.05 kJ/kg · K.

Product flows, kg/hr

	In	Out
Water	2,103	40
Sand	39,960	39,960
	——————+	——————+
Total	42,063	40,000

Evaporation 2,103 − 40 = 2,063 kg/hr
= 0.573 kg/sec

Net heat, kJ/hr

Net heat is the energy required by the process flow. Take the gas-exit and the product-exit temperature 90°C.

$$Q_1 = 2,063(2,504 + 1.886 \times 90 - 4.19 \times 10) = 5,429,486$$
$$Q_2 = 39,960(90 - 10)0.840 = 2,685,312$$
$$Q_3 = 40(90 - 10)4.19 = \underline{\quad 13,408 \quad} +$$
$$Qtot_1 = 8,128,206$$

$Qtot_1$ is the heat needed by the process flow which is transferred by the drying gas that is cooled down from 750 to 90°C. In addition, the drying gas supplies heat for convection and radiation losses. A factor of 1.1 appears to be satisfactory for steady-state operation.

$$Qtot_2 = 8,128,206 \times \frac{750 - 10}{750 - 90} \times 1.1 = 10,024,787 \text{ kJ/hr}$$

This amount of heat is transferred to the drying gas by the combustion of natural gas.

$$\frac{10,024,787}{2,063} = 4,859 \text{ kJ/kg of water evaporated}$$

Figure 5.11 High capacity fluid bed (center) with flash dryer (left) as predrying stage and product cooling in fluid bed (center-right). Product: polypropylene. (Courtesy of Niro Atomizer, Soeborg, Denmark)

The long-term consumption figure (start-up, shut-down, cleaning) is probably a factor 1.2 higher.

1.2 × 4,859 = 5,831 kJ/kg of water

Fluidizing air blower

Fluid bed height is 0.5 m. Bulk density of sand is 1,500 kg/m^3. Pressure drop in the fluidized bed is 1,500 × 9.81 × 0.5 = 7,357.5 N/m^2. Pressure drop of the air distributor is about 0.25 × 7,357.5 = 1,839.4 N/m^2.

7,357.5 + 1,839.4 + 2,500 = 11,697 N/m^2

2,500 N/m^2: combustion chamber's Δp.

Air quantity:

$$\frac{10{,}024{,}787}{1.05 \times (750 - 10)} = 12{,}902 \text{ kg/hr}$$

Air volume:

$$\frac{12{,}902}{1.29} \times \frac{283}{273} = 10{,}368 \text{ m}^3\text{/hr at } 10°C$$

Fan power consumption:

$$\frac{10{,}368 \times 11{,}697}{3{,}600 \times 1{,}000 \times 0.5} = 67.4 \text{ kW}$$

Install an 80-kW motor. This power consumption also includes the combustion air blower.

Exhaust fan

The exhaust fan draws the warm, moisture-laden air into the cyclones. Cyclone pressure drop is $1{,}500 \text{ N/m}^2$.

Air density at 90°C:

$$1.29 \times \frac{273}{273+90} = 0.97 \text{ kg/m}^3.$$

Airflow in the freeboard section:

$$1.1 \times \frac{12{,}902}{0.97} = 14{,}631 \text{ m}^3\text{/hr.}$$

The factor 1.1 is used to account for the attraction of ingress air. Water vapor flow in the freeboard section:

$$\frac{2{,}063}{0.60} = 3{,}438 \text{ m}^3\text{/hr.}$$

Water vapor density 0.60 kg/m^3.
Total gas flow in the freeboard section $14{,}631 + 3{,}438 = 18{,}069 \text{ m}^3\text{/hr.}$
Fan power consumption:

$$\frac{18{,}069 \times 1{,}500}{3{,}600 \times 1{,}000 \times 0.5} = 15.1 \text{ kW}$$

Install a 20-kW motor.

Fluid-bed dryer diameter

To avoid excessive dust entrainment, a superficial gas velocity just above the fluidized bed of 0.8 m/sec is taken. This velocity is reduced to 0.4 m/sec in a disengaging section.

$$\frac{\pi}{4} \times D^2 \times 0.8 = \frac{18,069}{3,600}$$

hence

$D = 2.83$ m

Diameter of the disengaging section $\sqrt{2} \times 2.83 = 4.00$ m

Miscellaneous power consumption
E.g., rotary locks.
Take 10 kW.

Consumption figures
1 kWhr costs $0.06.
1 nm^3 of natural gas supplies 32,000 kJ.
1 nm^3 of natural gas costs $0.1.
Natural gas costs:

$$\frac{5,831 \times 2,063 \times 0.1}{40 \times 32,000} = \$0.94 \text{ per ton of sand}$$

Electricity costs:

$$\frac{(67.4 + 15.1 + 10)1.2 \times 0.06}{40} = \$0.17 \text{ per ton of sand}$$

Factor 1.2: to arrive at the long-term consumption figure.

Remarks
The relative humidity of the exhaust gas may be checked. Total amount of air is $1.1 \times 12,902 = 14,192$ kg/hr. This air is not dry. Sixty percent relative humidity at 10°C means 0.005 kg of water vapor per kg of dry air.

Dry air   14,121
Water vapor   71
————— +
14,192 kg/hr

Evaporation   2,063 kg/hr

A third source of water vapor is the combustion of natural gas.

Natural gas consumption:

$$\frac{4,859 \times 2,063}{32,000} = 313.3 \text{ nm}^3/\text{hr}$$

Water formation is approximately 425 kg/hr.

71 + 2,063 + 425 = 2,559 kg/hr

2,559/14,121 = 0.181 kg of water vapor/kg of dry air.

Molar ratio is 0.225.

Water vapor partial pressure is 0.225 bar

Saturated water vapor pressure at 90°C is 0.7 bar

Relative humidity

$$\frac{0.225}{0.7} \times 100 = 32\%$$

5.6 Computer program (rectangular dryer)

The computer program in BASIC can be found in Appendix 5.1. Much of it can be comprehended directly by referring to Section 4.2 in Chapter 4. The additional remarks that follow are offered to aid comprehension:

1. Lines 60 through 180 deal with the input.
2. Lines 200 through 240 concern the mass balance for the process flow (see Section 4.2).
3. A FOR-NEXT loop starts on line 270. The variable cost and the entering and leaving gas flows are automatically calculated for TA_{IN} = 100, 200, 300, . . . °C up to a programmed T_{MAX}. Line 880 contains the statement to redo the loop.
4. Line 290 relates the drying-gas outlet temperature to the drying-gas inlet temperature; Figure 5.12 shows this relationship, with data taken from the literature (Williams-Gardner, 1971; Tailor, 1969) and data recorded in Akzo plants. The relationship is: TA_{OUT} = 0.083 × TA_{IN} + 48.3 °C. Fourteen sets of data were taken into account. The relationship was established by means of linear regression. The correlation coefficient is 0.930. Note that accepting a TA_{OUT}-value by employing line 290 can, at low-temperature drying, lead to considerable discrepancies. Low-temperature drying is defined as drying with an air-inlet temperature approximately between 40 and 100°C. For example, the relation predicts TA_{OUT} = 56.6°C, when TA_{IN} = 100°C. The useful temperature drop is 100 − 56.6 = 43.4°C.
      If, in actual fact, TA_{OUT} = 45°C, when TA_{IN} = 100°C, the useful temperature drop is considerably larger, i.e., 55°C. The useful temperature drop is important for the dryer design.
5. Lines 300 through 330 regard the process-stream enthalpy change. It is assumed that the drying-gas outlet temperature and the product temperature are equal.

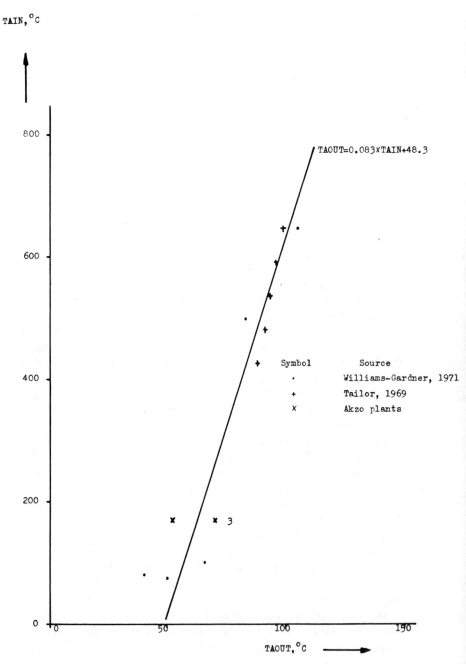

Figure 5.12 Relationship between air-inlet and air-outlet temperatures for rectangular fluid-bed dryers.

6. Line 340 calculates the amount of heat to be put into the drying-gas flow, with 10°C taken as the average ambient temperature. The factor 1.15 allows for the steady-state heat losses. This factor can be calculated as an average value from measurements reported in the literature (Williams-Gardner, 1971). The implicit assumption is that the energy consumptions reported were established by means of a measuring period of, e.g., 12 hr. It is assumed that the reported figures do not concern long-term energy consumption (see Section 4.2).

    The data on 3 out of 5 rectangular fluid-bed dryers were interpreted regarding the primary energy consumption and the energy consumption predicted by the model developed for the computer program, with 2 out of 5 sets of data disregarded because of low air-inlet temperatures (80 and 75°C). Two slightly different calculation methods were used:

    Starting points for both methods:

    TA_{IN} as indicated

    $TP_{OUT} = TA_{OUT}$

    $TP_{IN} = 20°C$

    Ambient temperature 10°C

    Starting point Method No. 1: $TA_{OUT} = 0.083 \times TA_{IN} + 48.3$
    Starting point Method No. 2: TA_{OUT} as indicated
    Method No. 1: ratios, which indicate consumption/calculated consumption, are 1.22, 1.02, and 1.29 (arithmetic average 1.18). The factor 1.15 was not used for the calculation of the consumption.
    Method No. 2: ratios, which indicate consumption/calculated consumption, are 1.07, 1.08, and 1.27 (arithmetic average 1.14). The selection of 1.15 as the factor for the steady-state heat losses is in line with experience within Akzo. It is in line with consumption figures reported for circular fluid-bed dryers also (Williams-Gardner, 1971).

7. Calculation of the dryer's size starts on line 350.
8. Line 360 concerns the air specific mass at TA_{OUT}.
9. Line 370 calculates the volumetric flow of the spent drying air. The factor 1.1 allows for the attraction of ingress air. In lines 380, 390, and 400 the evaporated water volumetric flow is added to the airflow. The relation of line 370 is approximately correct for both steam heating and direct heating. The specific heat 1.05 kJ/kg · K is a percentage too high for steam heating (air specific heat at 0°C is 1.0 kJ/kg · K and at 300°C, 1.05 kJ/kg · K).

    For direct heating, the value of 1.05 kJ/kg · K is about 4% too low for combustion gases cooling down from 750°C by water pick-up in the dryer. However, the specific mass prediction is approximately 2% too high in that case.

10. Lines 410 through 440 calculate the area.

11. Lines 450 through 470 serve to arrive at a gas exhaust flow (multiple of 1,000 m³/hr), which is useful in requesting a quotation.

12. The heating medium cost is calculated in lines 490 and 500. The multiplication with *Lofa* (the load factor) allows for the extra heat losses in the long run. *Lofa* is introduced in line 70; the background is explained in Section 4.2. The given numerical values are based on actual measurements in factories.

13. Line 510 gives the efficiency in kJ/kg of evaporated water. The power cost is the next item to be calculated.

14. Line 530 calculates the fluidizing air volumetric flow at 10°C, which relationship is approximately correct for steam heating and direct heating. For steam heating the comments made for line 370 are applicable. For direct heating, the value 1.05 kJ/kg · K is about 4% too low for combustion gases cooling down from 750°C by water pick-up in the dryer. Moreover, this relationship also includes the heating-medium flow. Adding up these two effects leads to the prediction being approximately 5% too high if TA_{IN} = 750°C (conservative).

15. Lines 540 through 560 calculate the air inlet flow (as a multiple of 1,000 m³/hr), useful for requesting a quotation.

16. Line 570 calculates the pressure loss due to the fluidized bed.

17. Line 580 states that the distributor's pressure loss is about one-half the bed's pressure loss. This is not an unrealistic situation for stable shallow fluidized beds (see Section 5.3).

17. The total Δp for the fluidizing air fan is arrived at via line 590, with 2,500 N/m² being added for a combustion chamber. A line burner and a steam heater cause smaller pressure losses, whereas there are combustion chambers with higher pressure losses: 2,500 N/m² (250 mm wg) is considered an approximation. The possible deviation is not too serious because the relation is used to calculate part of the electricity consumption. The electricity consumption *per se* is of second-order importance compared to the heating medium consumption.

   The simple model is approximately in line with literature data (Williams-Gardner, 1971) and plant measurements.

18. The exhaust fan's power consumption is calculated in line 610. A pressure drop of 1,500 N/m² across a cyclone or baghouse is taken.

19. Line 620 calculates the total power consumption, with 15 kW added to allow for vibration, rotary locks, etc.

20. Line 650 adds the cost elements.

The remainder of the program concerns the output. The program itself is followed by an example. It regards the continuous drying of vacuum-pan salt. Calculations were carried out from

$TA_{IN} = 100$ through $TA_{IN} = 800°C$.
Equipment used: IBM PC XT and printer Epson FX-80.

5.7 Investment aspects

Carrier Vibrating Equipment, Inc., kindly provided budget prices for vibrating rectangular fluid-bed dryers. These prices do not include fans, ducts, heater, and dust-separation equipment. The prices were provided in 1986.

Product: Drivert sugar
TA_{IN}: 71°C
TP_{OUT}: 32°C
A_1: 3% by wt
A_2: 0.4% by wt
ρ_b: 640 kg/m³

Parts coming into contact with the process-stream stainless steel AISI 304.

	Dryer No. 1	Dryer No. 2
Capacity, kg/hr	3,180–4,540	6,360–9,080
Width, m	1.22	—
Length, m	6.25	—
Drying area, m²	7.6	
Price (US)	$50,000	$78,000

The capacity of Dryer No. 2 is twice that of Dryer No. 1. These two prices are in line with the 0.65-rule, i.e., $\text{Price}_1 = \text{Price}_2 (\text{Capacity ratio})^{0.65}$.

5.8 Notation

A_1	Feed moisture content	% by wt
A_2	Product moisture content	% by wt
Air_1	Airflow taken in by the dryer	m³/hr
Air_2	Airflow taken in by the dryer	
Air_3	Airflow taken in by the dryer	
$Airflo_1$	Airflow taken in by the dryer	m³/hr
$Airflo_2$	Airflow leaving the dryer	m³/hr

$Area$	Fluid-bed area	m^2
B	Bed width	m
Cal_{VAL}	Fuel or steam calorific value	mJ/t
		mJ/nm^3
Cap	Product-mass flow	kg/hr
CP	Solid specific heat	kJ/kg · K
C_p	Air specific heat	kJ/kg · K
D	Fluid-bed diameter	m
Dp_1	Heater, plate, and bed pressure loss	N/m^2
Dp_{BED}	Fluidized-bed pressure loss	N/m^2
Dp_{PLATE}	Distributor plate pressure loss	N/m^2
d_p	Particle size	μm
$\underline{d}_p$	Average particle size	μm
d_{sv}	Mean surface/volume diameter of particles	μm
dx	Small bed length	m
$dT_p(x)$	Small product-temperature increase	K
$Effic$	Dryer efficiency	kJ/kg
$Evap$	Dryer water evaporation load	kg/hr
F_1	Fluid-bed area	m^2
F_2	Fluid-bed area	
F_3	Fluid-bed area	
Gas_2	Gas flow leaving the dryer	m^3/hr
Gas_3	Gas flow leaving the dryer	
Gas_4	Gas flow leaving the dryer	
$Gasflo_2$	Gas flow leaving the dryer	m^3/hr
G_{mf}	Minimum fluidization mass velocity	kg/m^2 · sec
g	Acceleration due to gravity	m/s^2
H	Enthalpy/Settled bed height	kJ/kg, m
h	Settled bed height	m
	Integer in FOR-NEXT loop	
h_{exp}	Expanded bed height	m
L_1	Bed length for free-water evaporation	m
L_2	Bed length for crossflow heating	m
L_3	Bed length for crossflow cooling	m
$Lofa$	Load factor	g
$Medco$	Medium cost	\$/nm^3
		\$/t
$Medcos$	Medium cost per metric ton of product	\$/t
$Medflo$	Medium flow	nm^3/hr
		t/hr
P	Power consumption	kW
Pow	Power consumption	kW

Pow_{CO}	Power cost	\$/kW
Pow_{COS}	Power cost per metric ton of product	\$/t
Pow_{G1}	Fan power required for Air_1	kW
Pow_{G2}	Fan power required for Gas_2	kW
Q_1	Heat flow for the evaporated water	kJ/hr
Q_2	Heat flow for the dry solids	kJ/hr
Q_3	Heat flow for the residual water	kJ/hr
$Qtot_1$	Net heat	kJ/hr
$Qtot_2$	Steady-state dryer-heat requirement	kJ/hr
R_A	Exhaust-air specific mass	kg/m^3
R_B	Bulk density solid	kg/m^3
R_W	Water vapor specific mass	kg/m^3
Re	Reynolds number	
Sol	Dry-solids flow	kg/hr
TA_{IN}	Drying air temperature	°C
TA_{OUT}	Spent drying-gas temperature	°C
Tf	Feed temperature	°C
T_{MAX}	Maximum drying air temperature	°C
TP_{IN}	Feed temperature	°C
TP_{OUT}	Dried product temperature	°C
$T_p(x)$	Product temperature at position x	°C
$T_p(x = 0)$	Initial product temperature	°C
$T_p(x = L_2)$	Final product temperature	°C
U	Linear gas velocity just above fluid bed	m/sec
Var_{COS}	Variable cost per metric ton or product	\$/t
v_{mf}	Minimum fluidization velocity	m/sec
v_F	Fluidization velocity	m/sec
WA_{FLO}	Evaporated water flow	m^3/hr
W_{IN}	Water flow to the dryer	kg/hr
W_{OUT}	Water flow leaving the dryer	kg/hr
X	Weight fraction	
x	Length	m
y	Air water content	kg/kg
y_o	Leaving air water content	kg/kg
y_i	Entering air water content	kg/kg
Δp	Pressure loss	N/m^2
μ	Fluidization gas viscosity	N · s/m^2
ϕ_{H_2O}	Water-vapor flow rate	kg/sec
ϕ_p	Product flow rate	kg/sec
ρ_b	Bulk density solid	kg/m^3
ρ_F	Fluidization gas specific mass	kg/m^3
ρ_S	Solid specific mass	kg/m^3

Appendix 5.1

```
LIST
10   WIDTH "LPT1:",132
20   LPRINT CHR$(15);
30   LPRINT "CONTINUOUS RECTANGULAR FLUID BED DRYER SCREENING PROGRAM"
40   LPRINT
50   LPRINT
60   INPUT "KG/H OUT";CAP
70   INPUT "LOAD FACTOR:FULL:1.1,NORMAL:1.3,LOW:1.5";LOFA
80   INPUT "FEED TEMPERATURE,DEGREES C";TF
90   INPUT "MAXIMUM DRYING AIR TEMPERATURE,h*100,h IS AN INTEGER,DEGREES C";TMAX
100  INPUT "MOISTURE CONTENT FEED,% BY WT";A1
110  INPUT "MOISTURE CONTENT PRODUCT,% BY WT";A2
120  INPUT "LINEAR GAS VELOCITY JUST ABOVE FLUIDIZED BED,M/S";U
130  INPUT "SOLID SPECIFIC HEAT,KJ/(KG*K)";CP
140  INPUT "BULK DENSITY SOLID,KG/M3";RB
150  INPUT "SETTLED BED HEIGHT,M";H
160  INPUT "MJ/METRIC T:STEAM,OIL;MJ/NM3:GAS";CALVAL
170  INPUT "U.S. $/METRIC T:STEAM,OIL;U.S.$/NM3:GAS";MEDCO
180  INPUT "POWER COST,U.S. $/KWH";POWCO
190  LPRINT
200  REM MASS BALANCE IN KG/H
210  WOUT=.01*CAP*A2
220  SOL=CAP-WOUT
230  WIN=SOL/(100-A1)*A1
240  EVAP=WIN-WOUT
250  LPRINT
260  REM STUDY OF THE DRYING AIR TEMPERATURE EFFECT
270  FOR TAIN=100 TO TMAX STEP 100
280    REM HEAT BALANCE IN KJ/H
290    TAOUT=.083*TAIN+48.3
300    Q1=EVAP*(2504+1.886*TAOUT-4.19*TF)
310    Q2=SOL*CP*(TAOUT-TF)
320    Q3=WOUT*4.19*(TAOUT-TF)
330    QTOT1=Q1+Q2+Q3
340    QTOT2=QTOT1*(TAIN-10)/(TAIN-TAOUT)*1.15
350    REM DRYER SIZING IN M
360    RA=355/(TAOUT+273)
370    AIRFLO2=QTOT2/(((TAIN-10)*1.05*RA)*1.1
380    RW=220/(TAOUT+273)
390    WAFLO=EVAP/RW
400    GAS2=AIRFLO2+WAFLO
410    F1=GAS2/(U*3600)
420    F2=F1/.1
430    F3=INT(F2)+1
440    AREA=.1*F3
450    GAS3=GAS2/1000
460    GAS4=INT(GAS3)+1
470    GASFLO2=GAS4*1000
480    REM HEATING MEDIUM COST IN U.S. $ PER METRIC T
490    MEDFLO=QTOT2/(CALVAL*1000)
500    MEDCOS=MEDFLO*MEDCO/CAP*1000*LOFA
510    EFFIC=MEDFLO*CALVAL*1000/EVAP*LOFA
520    REM POWER COST IN U.S. $ PER METRIC T
530    AIR1=QTOT2/(((TAIN-10)*1.05*1.24)
540    AIR2=AIR1/1000
550    AIR3=INT(AIR2)+1
560    AIRFLO1=AIR3*1000
570    DPBED=RB*H*9.810001
```

```
580    DPPLATE=.5*DPBED
590    DP1=DPBED+DPPLATE+2500
600    POWG1=AIR1*DP1/(3600*.5*1000)
610    POWG2=GAS2*1500/(3600*.5*1000)
620    POW=POWG1+POWG2+15
630    POWCOS=POW*POWCO/CAP*1000*LOFA
640    REM ADDING COSTS IN U.S. $ PER METRIC T
650    VARCOS=MEDCOS+POWCOS
660    REM OUTPUT
670    LPRINT
680    LPRINT
690    LPRINT
700    LPRINT USING "DRYING AIR TEMPERATURE,DEGREES C                  ####";TAIN
710    LPRINT USING "EXHAUST TEMPERATURE,DEGREES C                     ###";TAOUT
720    LPRINT USING "EVAPORATION,KG/H                                  ####";EVAP
730    LPRINT
740    LPRINT USING "EFFICIENCY,KJ/KG EVAPORATED                       #####";EFFIC
750    LPRINT
760    LPRINT USING "AREA,M2                                           ##.#";AREA
770    LPRINT
780    LPRINT USING "MEDIUM COST,U.S. $/METRIC T OF PRODUCT            ###.##";MEDCOS
790    LPRINT
800    LPRINT USING "POWER COST,U.S. $/METRIC T OF PRODUCT             ##.##";POWCOS
810    LPRINT
820    LPRINT USING "RELEVANT VARIABLE COSTS,U.S. $/METRIC T OF PRODUCT ###.##";VARCOS
830    LPRINT
840    LPRINT USING "INLET AIR FLOW AT 10 DEGREES C,M3/H               ######";AIRFLO1
850    LPRINT
860    LPRINT USING "EXHAUST GAS FLOW AT TAOUT,M3/H                    ######";GASFLO2
870    LPRINT
880    NEXT TAIN
890 LPRINT
900 WIDTH "LPT1:",80
910 LPRINT CHR$(18)
920 END
D


RUN
CONTINUOUS RECTANGULAR FLUID BED DRYER SCREENING PROGRAM


KG/H OUT? 20000
LOAD FACTOR:FULL:1.1,NORMAL:1.3,LOW:1.5? 1.3
FEED TEMPERATURE,DEGREES C? 20
MAXIMUM DRYING AIR TEMPERATURE,h*100,h IS AN INTEGER,DEGREES C? 800
MOISTURE CONTENT FEED,% BY WT? 3.0
MOISTURE CONTENT PRODUCT,% BY WT? .1
LINEAR GAS VELOCITY JUST ABOVE FLUIDIZED BED,M/S? 1.0
SOLID SPECIFIC HEAT,KJ/(KG*K)? .87
BULK DENSITY SOLID,KG/M3? 1250
SETTLED BED HEIGHT,M? .2
MJ/METRIC T:STEAM,OIL;MJ/NM3:GAS? 32
U.S. $/METRIC T:STEAM,OIL;U.S.$/NM3:GAS? .1
POWER COST,U.S. $/KWH? .06



DRYING AIR TEMPERATURE,DEGREES C              100
EXHAUST TEMPERATURE,DEGREES C                  57
```

```
EVAPORATION,KG/H                                         598

EFFICIENCY,KJ/KG EVAPORATED                            11149

AREA,M2                                                  15.7

MEDIUM COST,U.S. $/METRIC T OF PRODUCT                   1.04

POWER COST,U.S. $/METRIC T OF PRODUCT                    0.83

RELEVANT VARIABLE COSTS,U.S. $/METRIC T OF PRODUCT       1.87

INLET AIR FLOW AT 10 DEGREES C,M3/H                    44000

EXHAUST GAS FLOW AT TAOUT,M3/H                         57000



DRYING AIR TEMPERATURE,DEGREES C                        200
EXHAUST TEMPERATURE,DEGREES C                            65
EVAPORATION,KG/H                                        598

EFFICIENCY,KJ/KG EVAPORATED                            8103

AREA,M2                                                  5.7

MEDIUM COST,U.S. $/METRIC T OF PRODUCT                   0.76

POWER COST,U.S. $/METRIC T OF PRODUCT                    0.33

RELEVANT VARIABLE COSTS,U.S. $/METRIC T OF PRODUCT       1.08

INLET AIR FLOW AT 10 DEGREES C,M3/H                    16000

EXHAUST GAS FLOW AT TAOUT,M3/H                         21000



DRYING AIR TEMPERATURE,DEGREES C                        300
EXHAUST TEMPERATURE,DEGREES C                            73
EVAPORATION,KG/H                                        598

EFFICIENCY,KJ/KG EVAPORATED                            7861

AREA,M2                                                  3.8

MEDIUM COST,U.S. $/METRIC T OF PRODUCT                   0.73

POWER COST,U.S. $/METRIC T OF PRODUCT                    0.23

RELEVANT VARIABLE COSTS,U.S. $/METRIC T OF PRODUCT       0.97

INLET AIR FLOW AT 10 DEGREES C,M3/H                    10000

EXHAUST GAS FLOW AT TAOUT,M3/H                         14000
```

```
DRYING AIR TEMPERATURE,DEGREES C                      400
EXHAUST TEMPERATURE,DEGREES C                          82
EVAPORATION,KG/H                                      598

EFFICIENCY,KJ/KG EVAPORATED                          8000

AREA,M2                                                3.1

MEDIUM COST,U.S. $/METRIC T OF PRODUCT                 0.75

POWER COST,U.S. $/METRIC T OF PRODUCT                  0.19

RELEVANT VARIABLE COSTS,U.S. $/METRIC T OF PRODUCT     0.94

INLET AIR FLOW AT 10 DEGREES C,M3/H                  8000

EXHAUST GAS FLOW AT TAOUT,M3/H                      11000
```

```
DRYING AIR TEMPERATURE,DEGREES C                      500
EXHAUST TEMPERATURE,DEGREES C                          90
EVAPORATION,KG/H                                      598

EFFICIENCY,KJ/KG EVAPORATED                          8266

AREA,M2                                                2.6

MEDIUM COST,U.S. $/METRIC T OF PRODUCT                 0.77

POWER COST,U.S. $/METRIC T OF PRODUCT                  0.17

RELEVANT VARIABLE COSTS,U.S. $/METRIC T OF PRODUCT     0.94

INLET AIR FLOW AT 10 DEGREES C,M3/H                  6000

EXHAUST GAS FLOW AT TAOUT,M3/H                      10000
```

```
DRYING AIR TEMPERATURE,DEGREES C                      600
EXHAUST TEMPERATURE,DEGREES C                          98
EVAPORATION,KG/H                                      598

EFFICIENCY,KJ/KG EVAPORATED                          8588

AREA,M2                                                2.4

MEDIUM COST,U.S. $/METRIC T OF PRODUCT                 0.80

POWER COST,U.S. $/METRIC T OF PRODUCT                  0.15

RELEVANT VARIABLE COSTS,U.S. $/METRIC T OF PRODUCT     0.96

INLET AIR FLOW AT 10 DEGREES C,M3/H                  6000

EXHAUST GAS FLOW AT TAOUT,M3/H                       9000
```

DRYING AIR TEMPERATURE,DEGREES C	700
EXHAUST TEMPERATURE,DEGREES C	106
EVAPORATION,KG/H	598
EFFICIENCY,KJ/KG EVAPORATED	8940
AREA,M2	2.2
MEDIUM COST,U.S. $/METRIC T OF PRODUCT	0.84
POWER COST,U.S. $/METRIC T OF PRODUCT	0.14
RELEVANT VARIABLE COSTS,U.S. $/METRIC T OF PRODUCT	0.98
INLET AIR FLOW AT 10 DEGREES C,M3/H	5000
EXHAUST GAS FLOW AT TAOUT,M3/H	8000

DRYING AIR TEMPERATURE,DEGREES C	800
EXHAUST TEMPERATURE,DEGREES C	115
EVAPORATION,KG/H	598
EFFICIENCY,KJ/KG EVAPORATED	9311
AREA,M2	2.1
MEDIUM COST,U.S. $/METRIC T OF PRODUCT	0.87
POWER COST,U.S. $/METRIC T OF PRODUCT	0.14
RELEVANT VARIABLE COSTS,U.S. $/METRIC T OF PRODUCT	1.01
INLET AIR FLOW AT 10 DEGREES C,M3/H	5000
EXHAUST GAS FLOW AT TAOUT,M3/H	8000

References

Baeyens, J. and D. Geldart (1974). Predictive calculations of flow parameters in gas fluidized beds and fluidization behaviour of various powders, *Fluidisation et ses applications, Compte-Rendu du Congrès International,* Cepadues-Editions, Toulouse, France.

Beeken, D.W. (1960) Thermodrying in fluidised beds, *Brit. Chem. Eng.,* 5, 484.

Leva, M. (1959). *Fluidization,* McGraw-Hill, New York.

Perry, R. H., D. W. Green, and J.O. Maloney (1984). *Perry's Chemical Engineers' Handbook,* McGraw-Hill, New York.

Tailor, J.P. (1969). Drying and bulk handling of potassium chloride, *Transactions of the Third Symposium on Salt,* Northern Ohio Geological Society, Inc., Cleveland.

Vaněcěk, V., M. Markvart, and R. Drbohlav (1966). *Fluidized Bed Drying,* Leonard Hill, London.

Viebrock, W. A. and A. E. Hodel (1983). Fluid-bed dryer eliminates calcining, reduces heat requirements by 7½%, *Chem. Proc.* (Chicago), 46, 20.

Vreeland, R. and J.A. Bacchetti, (1982). Equipment plugging eliminated with fluid bed dryer, *Chem. Proc.* (Chicago), 45, 24.

Williams-Gardner, A. (1971). *Industrial Drying,* Leonard Hill, London, England.

Winterstein, G., K. Rose, H. Viehweg, and L. Schreyer (1964). Cooling of particulate materials in a rectangular fluid bed, *Chem. Tech.,* 16, 106 (in German).

6

CONTINUOUS DIRECT-HEAT
ROTARY DRYING

6.1 Introduction

A general description of direct-heat rotary drying is provided in Section 6.2, followed by a discussion of the screening design methods in Section 6.3, and a worked example in Section 6.4.

Section 6.5 covers the development of a computer program that is incorporated in this volume as Appendix 6.1, and Section 6.6 deals with investment aspects.

6.2 General description

Essentially, in a direct-heat rotary dryer a continuous feed of wet particulate material is dried by contact with heated air, while being transported along the interior of a rotating cylinder (see Fig. 6.1), with the rotating shell acting as a conveying device and as a stirrer.

In addition, two other types of continuous rotary dryers exist: (a) the indirect-heat rotary dryer in which heat is transferred indirectly, e.g., a steam-tube dryer (to be treated in greater detail in Chapter 9), and (b) the indirect-direct rotary dryer, often considered a hybrid in which heat is transferred indirectly (by conduction and radiation) and then directly, as illustrated in Figure 6.2. The construction is more complicated than the direct-heat rotary dryer, but the heat losses to the surrounding area are minimized because the outer wall is not in contact with the hottest gas.

Generally, refractory-lined rotary kilns are used to create a heat treatment for a process stream and are not straightforward dryers. The hot gases

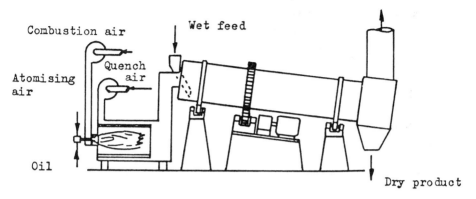

Figure 6.1 Typical cascading direct rotary dryer arranged for cocurrent operation. (From Nonhebel and Moss, 1971.)

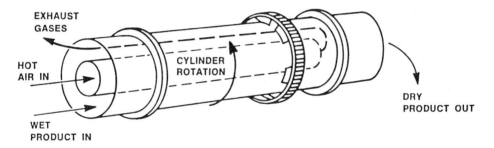

Figure 6.2 Indirect-direct rotary dryer. (Courtesy of C-E Raymond, Combustion Engineering, Inc., Chicago, Illinois.)

may enter at temperatures of up to 1650°C. This temperature range is also used in refractory-lined rotary incinerators.

Airswept rotary coolers may be used to cool a product to an appropriate temperature for bagging and storage. Rotary dryers that also have a cooling element exist: drying occurs cocurrently in the first section and cooling is performed countercurrently in the second. The spent gases can be combined and exhausted at, e.g., two-thirds of the shell length. Rotary dryers with coolers are used to process wet sugar from a centrifuge.

Direct-heat rotary dryers can be classified according to their relationship between product and gas flow: (a) cocurrent (parallel), (b) countercurrent, and (c) crossflow (Roto-Louvre dryers).

The Roto-Louvre dryer is a distinctive dryer in which the material to be processed rests on the bottom of a rotating cyclinder through which drying gas is passed that also passes through the bed (see Fig. 6.3). This dryer is suitable for friable material since the movement is gentle.

Direct-heat rotary dryers are universally applicable for particulate materials, but normally they cannot process solutions, slurries, or pastes. Transport and drying functions are independent of load over wide variations. The dryer needs little attention, with incrustation being prevented by the use of shell knockers (which are optional). Consequently, direct-heat rotary dryers are suitable for applications where supervision is either minimal or low grade. However, rotary dryers are mechanically complicated, with the running gear, seals, feed breechings, and discharge breechings requiring special attention. Maintenance costs per annum can reach 10% of the investment cost. Compare this to the 5% figure usually found for other convective dryers.

Air-velocity choice is important; a standard guide is to choose an air velocity providing a maximum carryover of 5–10% by weight of the dry solid. Probably 90% of all rotary dryers operate at gas velocities below 2 m/sec. The minimum particle size used in rotary dryers is approximately 100 μm.

Substantial hold-up is found in rotary dryers, with residence times between several minutes and one hour being common. Typical volumetric loadings vary between 10% and 15%. These factors have important implications for process safety. First, decomposition of the processed material can have serious consequences because of the large mass. Sugar beet pulp,

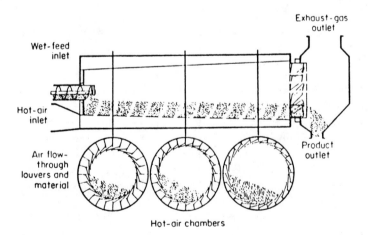

Figure 6.3   The Roto-Louvre dryer. (From Perry et al., 1984.)

e.g., is dried in direct-fired rotary dryers with gases entering at 800–900°C. If the operation is interrupted, the pulp is in stationary contact with the hot wall and may be incinerated. Second, another problem is associated with the evaporation of solvents. If there is a leakage of air into the dryer, the creation of explosive mixtures may result, whereas a reverse leakage may lead to a fire. However, solutions do exist because the rotary dryer is a mechanical engineer's dryer, so to speak.

Both the cocurrent and the countercurrent direct-heat rotary dryers are equipped with flights on the interior of the shell for lifting and showering the solids through the gas stream during passage. These lifting flights are offset every 0.6 to 2 m to ensure continuous and uniform curtains of solids in the gas chamber. The shape of the flights is chosen on the basis of the handling charcteristics of the solids (see Fig. 6.4). The cruciform flights reduce the falling height of the material and, hence, the entrainment; however, cleaning them can be a problem. In addition, to prevent entrainment, flights can be left out at the product end on cocurrent drying.

Diameter

The diameters of rotary dryers are generally between one and several meters. Any equipment exceeding this size range can create difficulties in

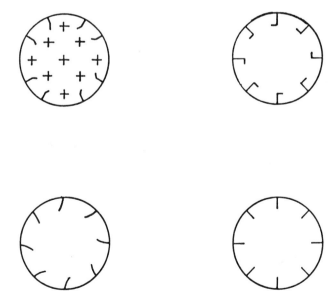

Figure 6.4 Possible direct-heat rotary dryer lifting-flight arrangements.

rail or road transport. On-site manufacturing can be an effective alternative.

Length

Usually, the length of a direct-heat rotary dryer is 5 to 8 times its diameter. For Roto-Louvre dryers: $2 \cdot D - 4 \cdot D$.

Slope

The slope of the cylinder from feed end to product end is usually between $0°$ and $- 5°$. However, positive slopes occur as well, these being used for cocurrent drying of relatively light materials.

Peripheral speed

Speed values between 0.1 and 0.5 m/sec are obtained, with speeds of 0.35 to 0.4 m/sec being quite common.

Number of flights

The flight count per circle is: $2.4 \cdot D - 3.0 \cdot D$ (Perry et al., 1984). D in feet.

Flight depth

$\frac{1}{12} \cdot D - \frac{1}{8} \cdot D$

In rotary dryers, dust separation is usually accomplished by cyclones. However, bag houses and wet scrubbers are also used. Figures 6.5, 6.6, and 6.7 illustrate several aspects of rotary dryers and rotary dryer operation.

Testing on a small-scale direct-heat rotary dryer will determine: (a) the gas inlet and outlet temperatures, (b) the feed- and product-moisture contents, (c) the permissible superficial gas velocity, and (d) the required residence time. The diameter of the small-scale dryer should not be less than 0.5 m.

6.3   Design methods

In this section, we deal with approximate design methods not specifically requiring test data. However, the availability of test data will produce more reliable designs. The methods outlined in this section can be used for this purpose, too.

The *mass balance* regarding the product is made first (see Chapter 4). The *drying-gas exit temperature* must be known to calculate the heat trans-

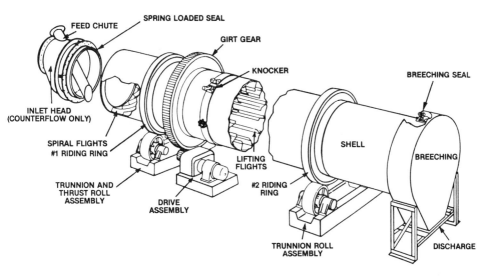

Figure 6.5 Component arrangements of a countercurrent direct-heat rotary dryer. (Courtesy of C-E Raymond, Combustion Engineering, Inc., Chicago, Illinois.)

Figure 6.6 Direct-heat rotary dryer during manufacture. (Courtesy of Swenson, Inc., Harvey, Illinois.)

Figure 6.7 Direct-heat rotary dryer installed in the user's plant. (Courtesy of Swenson, Inc., Harvey, Illinois.)

ferred to the product flow, $Qtot_1$. The proposed relationship between TA_{IN} and TA_{OUT} is

$$TA_{OUT} = 0.05 \cdot TA_{IN} + 64.5$$

Figure 6.8 contains a graphical presentation of plant measurements; the data were obtained from the literature (Williams-Gardner, 1971; Perry et al., 1984; Horgan, 1928) and observed for industrial dryers.

The data for counter and parallel airflow (11 points) were analyzed statistically. The relationship $TA_{OUT} = 0.0495 \times TA_{IN} + 63.76$ holds (linear regression), with the correlation coefficient being 0.916. Both constants are chosen higher to have a margin of safety.

A different approach was found in the literature (e.g., Perry et al., 1984).

$$N_t = \frac{TA_{IN} - TA_{OUT}}{(\Delta T)_m} \tag{6.1}$$

Empirical evidence indicates that direct-heat rotary dryers are most economically operated between $N_t = 1.5$ and $N_t = 2.5$ (N_t being the number of transfer units).

The true mean temperature difference between the hot gases and the material is $(\Delta T)_m$. When a considerable quantity of surface moisture is

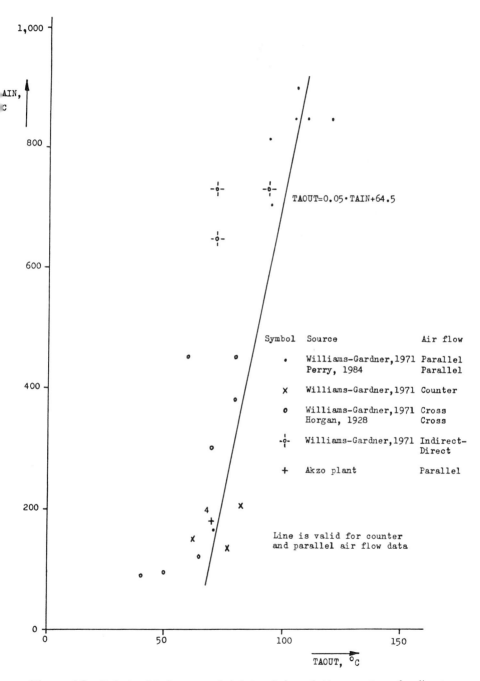

Figure 6.8 Relationship between air-inlet and air-outlet temperatures for direct-heat rotary dryers.

removed from the solids and the temperatures of the solids are unknown, a good approximation of $(\Delta T)_m$ is the logarithmic mean between the wet-bulb depressions of the drying air at the inlet and at the exit of the dryer.

The difference between these two methods cannot be neglected and therefore two worked examples are offered. We begin with the assumption that the solid is insoluble in water.

Example 1

$TA_{IN} = 145°C$
Ambient air temperature is 10°C
Relative humidity of the atmosphere is 60%
Heating is indirect by means of steam

Method 1

$TA_{OUT} = 0.05 \cdot 145 + 64.5 = 71.8°C$

Method 2

60% relative humidity at 10°C means 0.005 kg of vapor/kg of dry air. Adiabatic saturation temperature (numerically equals the wet-bulb temperature for the air/water system) of air at 145°C with 0.005 kg of water vapor/kg of dry air: 39°C (see Mollier's graph).

$$N_t = \log_e \left(\frac{TA_{IN} - T_w}{TA_{OUT} - T_w} \right)$$

N_t	1.5	2.0	2.5
TA_{OUT}, °C	62.7	53.3	47.7

Example 2

$TA_{IN} = 300°C$
Ambient air temperature is 10°C
Relative humidity of the atmosphere is 60%
Heating is direct by means of gas combustion

Method 1

$TA_{OUT} = 0.05 \cdot 300 + 64.5 = 79.5°C$

Method 2

The combustion of natural gas and the mixing of the hot gases with atmospheric air until 300°C is obtained, results in drying gas with approximately 0.025 kg of water vapor/kg of dry air. The corresponding adiabatic saturation temperature is 54°C.

N_t	1.5	2.0	2.5
TA_{OUT}, °C	108.9	87.3	74.2

It is assumed that the product-exit temperature equals TA_{OUT}, which is about right for cocurrent operation but may not be entirely correct for countercurrent operation. However, the effect on $Qtot_1$ is limited.

Now, $Qtot_1$ is multiplied by 1.25 to allow for the steady-state heat losses (see Chapter 4). This factor can be calculated as an average value from measurements reported in the literature (Williams-Gardner, 1971; Perry et al., 1984). These sources describe the steady-state heat consumption for the drying operation (Swenson, 1985; Buell, 1985). However, the literature data and plant measurements concerning very large industrial dryers (3 to 5 m diameter) indicate a value of 1.1 instead of 1.25, which is probably due to the attention paid to important industrial drying operations. Furthermore, the m^2/m^3 ratio is relatively small for large dryers. On the other hand, performance data for small dryers (e.g., 1-m diameter) in the literature point to values of up to 1.4. The drying airflow can be calculated by realizing that $1.25 \cdot Qtot_1$ is transferred by the air on cooling down from TA_{IN} to TA_{OUT}. This quantity is multiplied by 1.2 to allow for ingress air. To arrive at the total gas flow at the exit, the evaporated water-vapor flow must be added. Via an acceptable gas velocity, the diameter is obtained.

Again, an empirical check of the *length to diameter ratios* reported in the literature and found in plants gave values between 5 and 8. So, as a general rule, 7 can be taken, which is done in the computer program.

However, a different method to calculate the length is discussed in the literature (e.g., Perry et al., 1984). Drying in a direct-heat rotary dryer can be expressed as a heat-transfer mechanism as follows:

$$1.25 \cdot Qtot_1 = U \cdot a \cdot V (\Delta T)_m \qquad \text{kJ/hr} \qquad (6.2)$$

Miller et al. (1942), Friedman and Marshall (1949), and Seaman and Mitchell (1954) developed data for evaluating $U \cdot a$. The relationships can be reduced to:

$$U \cdot a = K \cdot G^n/D \qquad \text{kJ/m}^3 \cdot \text{hr} \cdot \text{K} \qquad (6.3)$$

McCormick (1962) compared the data and found that all experimental evidence can be described by $n = 0.67$. His comparison comprised: (a) experimental and commercial dryers, (b) countercurrent and cocurrent operation, (c) the drying and heating of different materials, (d) drying air temperatures up to at least 300°C, and (e) a maximum drying airflow of 4.4 kg/sec · m² dryer cross-section; however, 1 to 1.5 were more typical values.

K is a variable and depends on:

1. The physical properties of the material dried
2. The flight count
3. The flight depth
4. The flight load
5. The rotational speed

Perry et al. (1984) recommended $K = 0.5$ for dryers having a flight count of $2.4 \cdot D$ to $3.0 \cdot D$ and operating at shell peripheral speeds of 60 to 75 ft/min (D in ft). The American Institute of Chemical Engineers (AICHE) (1985) recommended $0.5 \leq K \leq 0.75$. K was found to approach 0.75 for high dryer loading (up to, e.g., 17%) and a large number of lifting flights:

$$1.25 \cdot Qtot_l = (0.5 \cdot G^{0.67}/D)V(\Delta T)_m$$
$$= 0.4 \cdot L \cdot D \cdot G^{0.67}(\Delta T)_m$$

However, the proportionality constants, 0.4 and 0.5, are applicable when using British units.

$$1.25 \cdot Qtot_l = 2.8 \cdot L \cdot D \cdot G^{0.67}(\Delta T)_m \quad \text{kJ/hr} \tag{6.4}$$

In kg/m² · hr, G is the mass flow of the gas. Perry et al. (1984) stated that air mass velocities in rotary dryers usually range from 0.5 to 5.0 kg/m² · sec. Experiments do show what should be the air rate with regard to dusting. An air rate of 1.4 kg/m² · sec can usually be safely used with 500-μm solids.

Once the air velocity is selected, we can then calculate D:

$$(\Delta T)_m = \frac{TA_{IN} - TA_{OUT}}{N_t}$$

So, equation (6.4) can be used to calculate the length. Still another point can be made after rearranging the equation (6.4):

$$L = \frac{1.25 \cdot Qtot_l \cdot N_t}{2.8 \cdot D \cdot G^{0.67}(TA_{IN} - TA_{OUT})}$$

$$L = \frac{\pi \cdot D^2 \cdot c_p \cdot G(TA_{IN} - TA_{OUT}) \cdot N_t}{4 \cdot 2.8 \cdot D \cdot G^{0.67}(TA_{IN} - TA_{OUT})}$$

$$\frac{L}{D} = 0.28 \cdot N_t \cdot c_p \cdot G^{0.33}$$

A value of 1.05 kJ/kg · K for c_p can be taken:

$$\frac{L}{D} = 0.29 \cdot N_t \cdot G^{0.33} \tag{6.5}$$

So, Equation (6.4) comes down to the fixation of an L/D-ratio for certain G- and N_t-values.

Matrix L/D. N_t \ G	1,800	3,600	7,200	10,800
1.5	5.2	6.5	8.2	9.3
2.0	6.9	8.6	10.9	12.4
2.5	8.6	10.8	13.6	15.5

$G = 3,600$ kg/m$^2 \cdot$ hr $= 1$ kg/m$^2 \cdot$ sec, thus, the $L/D -$ predictions of Perry et al. (1984) appear to be on the safe side when compared with practical values.

Friedman and Marshall (1949) investigated the *residence time* in a semi-technical rotary dryer. Perry et al. (1984) has provided relationships based on their work:

$$\tau = \frac{0.23 \cdot L}{S \cdot N^{0.9} \cdot D} \pm \frac{2 \cdot B \cdot L \cdot G}{F} \text{ min} \tag{6.6}$$

$$B = 5 \cdot \bar{d}_p^{-0.5}$$

The AICHE (1985) recommended the same relationship.

The plus sign refers to countercurrent flow and the negative sign to cocurrent operation. The second part to the right of Equation (6.6) is a correction for the airflow. Equation (6.6) can be used for air velocities not appreciably exceeding 1 m/sec.

Based on a review of the literature data (Williams-Gardner, 1971; Perry et al., 1984); private communications (Swenson, 1986), and measurements at the plant site, an empirical formula for the *motive power consumption* of a dryer with lifting flights is suggested:

$$Pow_{ROT} = 0.3 \times \pi/4 \times D^2 \times L \text{ kW} \tag{6.7}$$

However, lower values are found for slowly rotating dryers, e.g., sugar-beet pulp dryers.

The AICHE (1985) also recommended a relationship for the motive power consumption of a dryer with lifting flights.

$$Pow_{ROT} = N \left[(34.3 \times D \times w) + (1.39 \times (D + 0.6) \times W) + 0.73 \times W \right]/ 134,040 \text{ kW} \tag{6.8}$$

The expression $(D + 0.6)$ is meant to estimate the diameter of the cylinder riding ring. Furthermore, they assumed that the flights are sufficient to shower all the material in the cylinder.

6.4   Worked example

Product

General	15 metric ton/hr of mineral from the dryer
Temperature feed	15°C
Particle size	50% by wt larger than 600 μm
Initial moisture content	12% by wt (as fed)
Final moisture content	3% by wt (as discharged)
Solids, specific mass	2,250 kg/m^3
Solids, specific heat	0.8 kJ/kg · K
Solids, water solubility	nil
Dry solids, bulk density	1,400 kg/m^3

Process

Ambient temperature	10°C
Gas inlet temperature	700°C
Airflow	cocurrent
Exit gas velocity	1.5 m/sec

Water/air data

Water, specific heat	4.19 kJ/kg · K
Water vapor, specific heat	1.886 kJ/kg · K
Latent heat of evaporation at 0°C	2,504 kJ/kg
Water vapor, specific mass relationship at 10^5 Pa	220/(273 + T) kg/m^3
Air, specific heat (average)	1.05 kJ/kg · K
Air, specific mass relationship at 10^5 Pa	355/(273 + T) kg/m^3

It is assumed that pilot plant drying tests were not carried out. They should give a clue as to the permissible exit gas velocity. Furthermore, the required residence time is obtained. Thence, carry-over is assumed to be negligible, and product exit temperature is assumed to be equal to the gas exit temperature

Mass balance, kg/hr

	In	Out
Solids	14,550	14,550
Water	1,984 +	450 +
	16,534	15,000

$EVAP = 1,984 - 450 = 1,534$

Net heat, kJ/hr

$TA_{OUT} = 0.05 \cdot 700 + 64.5 = 99.5°C$

$$
\begin{aligned}
Q_1 &= 1{,}534(2{,}504 + 1.886 \cdot 99.5 - 15 \cdot 4.19) &&= 4{,}032{,}590 \\
Q_2 &= 14{,}550 \cdot 0.8(99.5 - 15) &&= 983{,}580 \\
Q_3 &= 450 \cdot 4.19(99.5 - 15) &&= 159{,}325 \\
\hline
Qtot_1 &&&= 5{,}175{,}495
\end{aligned}
$$

$$Qtot_2 = 1.25 \times \frac{(700 - 10)}{(700 - 99.5)} \times 5{,}175{,}495 = 7{,}433{,}579$$

Efficiency: $\dfrac{7{,}433{,}579}{1{,}534} = 4{,}846$ kJ/kg of water

Hot gas preparation unit

Burner capacity 7,433,579 kJ/hr. Install a burner with 35% spare capacity, i.e., a burner with a nominal capacity of 10,000 MJ/hr.

Air flow

$$\frac{7{,}433{,}579}{(700 - 10)1.05} = 10{,}260 \text{ kg/hr}$$

$R_A = 355/(273 + 15) = 1.233$ kg/m^3

$10{,}260/1.233 = 8{,}321$ m^3/hr

Fan

$\Delta p = 2{,}500$ N/m^2

$$Pow_{G1} = \frac{8{,}321 \cdot 2{,}500}{3{,}600 \cdot 1{,}000 \cdot 0.5} = 11.6 \text{ kW}$$

Take a 15-kW motor.

Dryer

The diameter calculation is based on outlet conditions:

$R_A = 355/(273 + 99.5) = 0.953$ kg/m^3

$R_W = 220/(273 + 99.5) = 0.591$ kg/m^3

$1.2 \times 10{,}260/0.953 = 12{,}919$ m^3/hr (airflow at the outlet including ingress air).

$WA_{FLO} = 1{,}534/0.591 = 2{,}596$ m^3/hr

$Gas_2 = 12{,}919 + 2{,}596 = 15{,}515$ m^3/hr

$\frac{\pi}{4}D^2 \cdot U \times 3{,}600 \times 0.85 = 15{,}515 \rightarrow D = 2.07$ m.

Take 2.10 m

$L = 7 \cdot 2.10 = 14.7$ m.

The factor 0.85 is used to compensate for flights, etc., occupying some of the cross-sectional area.

$$Pow_{ROT} = 0.3 \cdot \frac{\pi}{4} \cdot D^2 \cdot L = 15.3 \text{ kW}$$

Use a 40-kW motor for the rotation.

Residence time

$$B = 5 \cdot \bar{d}_p^{-0.5} = 5 \cdot 600^{-0.5} = 0.204$$

$$N = 4 \text{ rpm}$$

$$S = 0.010 \text{ (slope)}$$

$$F = \frac{15,000 \cdot 4}{\pi \cdot 2.1^2} = 4,331 \text{ kg/m}^2\cdot\text{hr}$$

$$\tau = \frac{0.23 \cdot 14.7}{0.010 \cdot 4^{0.9} \cdot 2.1} - \frac{2 \cdot 0.204 \cdot 14.7 \cdot 3,555}{4,331} = 46.2 - 4.9 = 41.3 \text{ min}$$

Volumetric loading

Solids volume in dryer $\dfrac{41.3}{60} \times \dfrac{15,000}{1,400} = 7.38 \text{ m}^3$

Volume % : $100 \times \dfrac{7.38 \cdot 4}{\pi \cdot 2.1^2 \cdot 14.7} = 14.5.$

Exhaust gas unit

Install a cyclone for 15,515 m³/hr, $\Delta p = 1,500 \text{ N/m}^2$.

$$Pow_{G2} = \frac{15,515 \cdot 1,500}{3,600 \cdot 1,000 \cdot 0.5} = 12.9 \text{ kW}$$

Install a 20-kW motor.

Consumption figures

Gas: $Cal_{VAL} = 32 \text{ MJ/nm}^3$
7,433.6/32 = 232 nm³/hr
232/15 = 15.5 nm³/ton of product
Lofa = 1.25
Long-run figure: $1.25 \cdot 15.5 = 19.4$, say, 20 nm³/ton

Electricity	Fan No. 1	11.6
	Fan No. 2	12.9
	Motive power	15.3
	Miscellaneous	15.0

54.8 kW; 3.7 kWhr/ton of product
Long-run figure: $1.25 \cdot 3.7 = 4.6 \text{ kWhr/ton}$

Costs

Guestimated price (see Table 6.1) $ 360,000
Investment 2.25 · 360,000 = $810,000
For an annual production of 8,000 × 15 = 120,000 tons, the fixed costs
are:

$$\frac{0.35 \cdot 810,000}{120,000} = \$2.36/t$$

Gas costs 20 · 0.1 = $2.00/t
Electricity costs 4.6 · 0.06 = $0.28/t

Note that several costs have not been taken into account, e.g., personnel
costs. These figures could be equal for different types of dryers.

6.5   Computer program

The computer program in BASIC is in Appendix 6.1. Reference is made to
the description of the program for the continuous rectangular fluid-bed
dryer (see Section 5.6; Section 4.2 may also be consulted).

Additional remarks

Input: lines 60–160.
Mass balance process flow: lines 180–220.
A FOR-NEXT loop exists between lines 250 and 860.
Line 270 relates the drying-gas exit temperature to the drying-gas inlet
  temperature: $TA_{OUT} = 0.05 \times TA_{IN} + 64.5$.
The observations regarding low-temperature drying made for fluid-bed
  drying also apply in this instance.
Process stream enthalpy change: lines 280–310.
The air-outlet temperature and the product-outlet temperature are as-
  sumed to be equal, which is about right for a cocurrent operation.
The ambient temperature is arbitrarily fixed at 10°C.
The factor 1.25 in line 320 allows for the steady-state heat losses. The
  origin of this factor is explained in Section 6.3.
Lines 350 through 380 calculate the exit gas flow in m³/hr.
The airflow is multiplied by 1.2 to allow for the attraction of ingress air.
The water vapor flow and the air are added up in line 380.
Lines 390 through 430 calculate the dryer size, $4*ATN(1) = \pi$.
The diameter is calculated as a multiple of one decimeter.
The length is selected as 7 times the diameter.
Lines 480 and 490 calculate the cost of the heating medium, the multipli-
  cation with *Lofa* (the load factor) allows for the increasing cost of the
  fuel medium per metric ton of product.

The background is explained in Section 4.2. The load factor is introduced into the program via line 70.

The power cost is the next item to be calculated. Line 520 calculates the
inlet airflow at 10°C, an average specific heat value of 1.05 kJ/kg · K is
taken.

Line 560 calculates the power required to pass the air through a gas
burner. To simplify matters, it is assumed that the pressure drop is
2,500 N/m² (255 mm wg).

The explanation is provided in Section 5.6. A steam heater would require about 50 mm wg only. However, the impact is not great because
one is dealing with a minor cost factor.

Line 570 states the power required to pass the air through the dryer and
the cyclone or the filter. A pressure loss of 150 mm wg is about
average. This figure can be derived from performance data reported
in the literature (Williams-Gardner, 1971; Perry et al., 1984) and is
consistent with plant measurements.

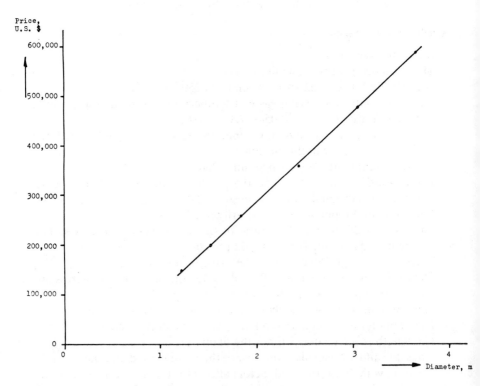

Figure 6.9 Relationship between price and diameter for stainless steel (304)
direct-heat rotary dryers. (Courtesy of Swenson, Inc., Harvey, Illinois.)

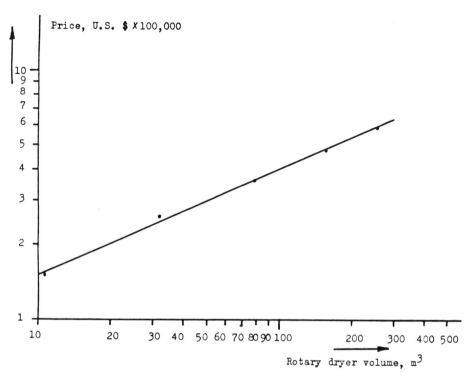

Figure 6.10 Relationship between volume and price for stainless steel (304) direct-heat rotary dryers. (Courtesy of Swenson, Inc., Harvey, Illinois.)

The motive power is found via an empirical formula:
$Pow_{ROT} = \text{ATN}(1) \cdot Diam^2 \cdot L \cdot .3$ (see Section 6.3).
Line 590 adds the power consumptions, an allowance of 15 kW is made,
    e.g., vibratory feeders, rotary locks, screws, and the like.

The remainder of the program concerns the output. Here, the instructions as to the desired output are given.

Appendix 6.1 contains the program. The program itself is followed by an example that examines the continuous drying of vacuum pan salt. Calculations are carried out from $TA_{IN} = 100$ through $TA_{OUT} = 800°C$. Notice that the costs per metric ton of product hardly vary from 300 to 800°C because the relatively large salt flow is heated to higher temperature values for higher TA_{IN}. However, when deciding the size of the cooler, this factor must be taken into account (see Fig. 6.11). Equipment used: IBM PC XT and printer Epson FX-80.

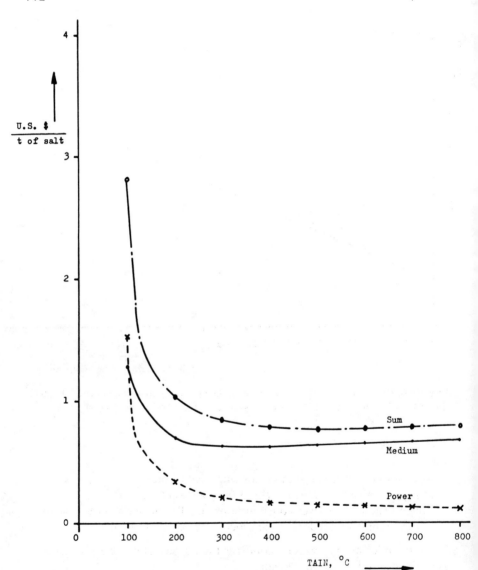

Figure 6.11   Rotary drying costs per metric ton of salt (see Appendix 6.1).

Table 6.1 Direct-Heat Countercurrent Rotary Dryer

Material   : anhydrous chemical salt
Feed condition   : 5% by wt of moisture, 50°C
Product condition   : 0% by wt of moisture, 150°C
Inlet air temperature : 650°C
Exit air temperature   : 120°C

Dryer size, m × m	1.22×9.14	1.52×10.67	1.83×12.19	2.44×16.76	3.05×21.34	3.66×24.38
Evaporation, kg/hr	204	318	454	816	1,270	1,814
Discharge, kg/hr	3,878	6,033	8,618	15,513	24,130	34,473
Work, 10^8 J/hr	9.556	14.90	21.27	38.22	59.49	84.97
Fuel (N.G.), m^3/hr	41.9	64.0	90.7	164.0	265.7	362.6
Exhaust volume, m^3/min	61.0	93.3	132.0	238.6	373.2	528.0
Exhaust fan, kW	3.75	5.63	7.50	15.0	22.5	30.0
Dryer drive, kW	5.63	11.3	18.8	56.3	93.8	150.0
Shipping wt, kg	12,500	17,700	25,000	35,000	73,000	100,000
Price FOB Chicago	$150,000	$200,000	$260,000	$360,000	$480,000	$590,000

Note: Assumed pressure drop in system: 200 mm. System includes air heater, dryer, drive, dust collector, ductwork, exhaust fan, and instrumentation. Prices are for 304 stainless-steel fabrication. For carbon steel, multiply given prices by 0.67.

Source: Courtesy of Swenson, Inc., Harvey, Illinois. 1986.

6.6   Investment aspects

The firm of Swenson, Inc., kindly provided prices for direct-heat counter-current rotary dryers (see Table 6.1). These data are also plotted in Figures 6.9 and 6.10. The correlation coefficient of the relationship $P = (180,691 \times Diam - 73,181)$ is almost 1. Figure 6.9 is a linear/linear-plot of the prices versus the diameter. Fig. 6.10 is a log/log-plot of the prices as a function of the dryer volume. The relationship $P = 56,754 \times V^{0.424}$ can be derived. The correlation coefficient of the relation $^{10}\log P = 0.424 \times {}^{10}\log V + 4.754$ is almost 1, too. The exponent 0.424 means that in going from 10-m^3 dryer volume to 250-m^3 dryer volume, the price increases relatively moderately. Exponential cost factors range from approximately 0.5 to about 0.8. In the case of an electrolytic cell hall complete with its auxiliaries the exponent might be 0.8. Capacity extension occurs via additional basic units in that case. An exponent of 0.5 is found, e.g., for evaporators; a larger capacity is provided by a larger version of the same type of equipment. Although the exponent is somewhat small, the Swenson data are in line with those above (see, e.g., The Institution of Chemical Engineers/The Association of Cost Engineers, 1977). The indications are valid for diameters up to 4 m. However, it should be realized that the plotting of the prices is done only as a function of the dryer's diameter, respectively, the dryer's volume and this simplifies the picture somewhat, as the prices do include the delivery of additional components (see Table 6.1).

Notation

A_1	Feed moisture content	% by wt
A_2	Product moisture content	% by wt
Air_1	Airflow taken in by the dryer	m^3/hr
Air_2	Airflow taken in by the dryer	
Air_3	Airflow taken in by the dryer	
$Airflo_1$	Airflow taken in by the dryer	m^3/hr
$Airflo_2$	Airflow leaving the dryer	m^3/hr
a	Contact area	m^2/m^3
B	$B = 5 \cdot \bar{d}_p^{-0.5}$	μm$^{-0.5}$
Cal_{VAL}	Fuel or steam calorific value	MJ/(nm^3 or t)
Cap	Product-mass flow	kg/hr
CP	Solid specific heat	kJ/kg $\cdot$ K
c_p	Air specific heat	kJ/kg $\cdot$ K
D	Dryer diameter	m or feet
D_1	Dryer diameter	m

D_2	Dryer diameter	
D_3	Dryer diameter	
Diam	Dryer diameter	m
$\bar{d}_p$	Average particle size	μm
Effic	Dryer efficiency	kJ/kg H_2O
Evap	Dryer water-evaporation load	kg/hr
F	Product mass rate	$kg/m^2 \cdot hr$
G	Gas (air) mass flow	$kg/m^2 \cdot hr$
Gas_2	Gas flow leaving the dryer	m^3/hr
Gas_3	Gas flow leaving the dryer	
Gas_4	Gas flow leaving the dryer	
$Gasflo_2$	Gas flow leaving the dryer	m^3/hr
h	Integer in FOR-NEXT loop	
K	Proportionality constant	$kJ/(m^{0.66} \cdot hr^{0.33} \cdot K \cdot kg^{0.67})$
L	Dryer length	m
Lofa	Load factor	
Medco	Medium cost	$\$/(nm^3$ or t)
Medcos	Medium cost per metric ton of product	$\$/t$
Medflo	Medium flow	$(nm^3$ or t)/hr
N	Rotational speed	min^{-1}
N_t	Number of transfer units	
n	Exponent in Equation (6.3)	
P	Price	\$
Pow	Power consumption	kW
Pow_{CO}	Power cost	\$/kWhr
Pow_{COS}	Power cost per metric ton of product	\$/t
Pow_{G1}	Fan power required for $Airflo_1$	kW
Pow_{G2}	Fan power required for Gas_2	kW
Pow_{ROT}	Motive power consumption	kW
Q_1	Heat flow to the evaporated water	kJ/hr
Q_2	Heat flow to the dry solids	kJ/hr
Q_3	Heat flow to the residual water	kJ/hr
$Qtot_1$	Net heat	kJ/hr
$Qtot_2$	Steady-state dryer-heat requirement	kJ/hr
R_A	Air specific mass	kg/m^3
R_W	Water vapor specific mass	kg/m^3
S	Slope of the dryer	

Sol	Dry-solids flow	kg/hr
T	Temperature	°C
TA_{IN}	Drying air temperature	°C
TA_{OUT}	Spent drying-gas temperature	°C
Tf	Feed temperature	°C
T_{MAX}	Maximum drying-air temperature	°C
T_w	Adiabatic saturation temperature	°C
U	Heat transfer coefficient	kJ/m² · hr · K
	Superficial air velocity out	m/sec
V	Dryer volume	m³
Var_{COS}	Relevant variable costs per metric ton of product	\$/t
W	Material plus dryer rotating weight	kg
w	Material rotating weight (hold-up)	kg
WA_{FLO}	Evaporated water flow	m³/hr
W_{IN}	Water flow to the dryer	kg/hr
W_{OUT}	Water flow leaving the dryer	kg/hr
$(\Delta T)_m$	True mean temperature difference	K
Δp	Pressure loss	N/m²
τ	Residence time	min

Appendix 6.1

```
LIST
10 WIDTH "LPT1:",132
20 LPRINT CHR$(15);
30 LPRINT "DIRECT-HEAT ROTARY DRYER SCREENING PROGRAM"
40 LPRINT
50 LPRINT
60 INPUT "KG/H OUT";CAP
70 INPUT "LOAD FACTOR;FULL:1.1,NORMAL:1.25,LOW:1.4";LOFA
80 INPUT "FEED TEMPERATURE,DEGREES C";TF
90 INPUT "MAXIMUM DRYING AIR TEMPERATURE,h*100,h IS AN INTEGER,DEGREES C";TMAX
100 INPUT "MOISTURE CONTENT FEED,% BY WT";A1
110 INPUT "MOISTURE CONTENT PRODUCT,% BY WT";A2
120 INPUT "SUPERFICIAL VELOCITY GAS OUT,M/S";U
130 INPUT "SOLID SPECIFIC HEAT,KJ/(KG*K)";CP
140 INPUT "MJ/METRIC T:STEAM,OIL;MJ/NM3:GAS";CALVAL
150 INPUT "U.S. $/METRIC T:STEAM,OIL;U.S. $/NM3:GAS";MEDCO
160 INPUT "POWER COST,U.S. $/KWH";POWCO
170 LPRINT
180 REM MASS BALANCE IN KG/H
190 WOUT=.01*CAP*A2
200 SOL=CAP-WOUT
```

```
210  WIN=SOL/(100-A1)*A1
220  EVAP=WIN-WOUT
230  LPRINT
240  REM STUDY OF THE DRYING AIR TEMPERATURE EFFECT
250  FOR TAIN=100 TO TMAX STEP 100
260    REM HEAT BALANCE IN KJ/H
270    TAOUT=.05*TAIN+64.5
280    Q1=EVAP*(2504+1.886*TAOUT-4.19*TF)
290    Q2=SOL*CP*(TAOUT-TF)
300    Q3=WOUT*4.19*(TAOUT-TF)
310    QTOT1=Q1+Q2+Q3
320    QTOT2=QTOT1*(TAIN-10)/(TAIN-TAOUT)*1.25
330    REM DRYER SIZING IN M
340    RA=355/(TAOUT+273)
350    AIRFLO2=QTOT2/((TAIN-10)*1.05*RA)*1.2
360    RW=220/(TAOUT+273)
370    WAFLO=EVAP/RW
380    GAS2=AIRFLO2+WAFLO
390    D1=SQR(GAS2/(U*ATN(1)*3600*.85))
400    D2=D1/.1
410    D3=INT(D2)+1
420    DIAM=.1*D3
430    L=7*DIAM
440    GAS3=GAS2/1000
450    GAS4=INT(GAS3)+1
460    GASFLO2=GAS4*1000
470    REM HEATING MEDIUM COST IN U.S. $ PER METRIC T
480    MEDFLO=QTOT2/(CALVAL*1000)
490    MEDCOS=MEDFLO*MEDCO/CAP*1000*LOFA
500    EFFIC=MEDFLO*CALVAL*1000/EVAP*LOFA
510    REM POWER COST IN U.S. $ PER METRIC T
520    AIR1=QTOT2/((TAIN-10)*1.05*1.24)
530    AIR2=AIR1/1000
540    AIR3=INT(AIR2)+1
550    AIRFLO1=AIR3*1000
560    POWG1=AIR1*2500/(3600*.5*1000)
570    POWG2=GAS2*1500/(3600*.5*1000)
580    POWROT=ATN(1)*DIAM^2*L*.3
590    POW=POWG1+POWG2+POWROT+15
600    POWCOS=POW*POWCO/CAP*1000*LOFA
610    REM ADDING COSTS IN U.S. $ PER METRIC T
620    VARCOS=MEDCOS+POWCOS
630    REM OUTPUT
640    LPRINT
650    LPRINT
660    LPRINT
670    LPRINT USING "DRYING AIR TEMPERATURE,DEGREES C                    ####";TAIN
680    LPRINT USING "EXHAUST GAS TEMPERATURE,DEGREES C                   ###";TAOUT
690    LPRINT USING "EVAPORATION,KG/H                                    ####";EVAP
700    LPRINT
710    LPRINT USING "EFFICIENCY,KJ/KG EVAPORATED                        #####";EFFIC
720    LPRINT
730    LPRINT USING "DIAMETER,M                                         #.##";DIAM
740    LPRINT USING "LENGTH,M                                          ##.##";L
750    LPRINT
760    LPRINT USING "MEDIUM COST,U.S. $/METRIC T OF PRODUCT            ###.##";MEDCOS
770    LPRINT
780    LPRINT USING "POWER COST,U.S. $/METRIC T OF PRODUCT              ##.##";POWCOS
790    LPRINT
800    LPRINT USING "RELEVANT VARIABLE COSTS,U.S. $/METRIC T OF PRODUCT ###.##";VARCOS
810    LPRINT
```

```
820   LPRINT USING "INLET AIR FLOW AT 10 DEGREES C,M3/H          ######";AIRFLO1
830   LPRINT
840   LPRINT USING "EXHAUST GAS FLOW AT TAOUT,M3/H               ######";GASFLO2
850   LPRINT
860 NEXT TAIN
870 LPRINT
880 WIDTH "LPT1:",80
890 LPRINT CHR$(18)
900 END
O
```

```
RUN
DIRECT-HEAT ROTARY DRYER SCREENING PROGRAM


KG/H OUT? 20000
LOAD FACTOR;FULL:1.1,NORMAL:1.25,LOW:1.4? 1.25
FEED TEMPERATURE,DEGREES C? 50
MAXIMUM DRYING AIR TEMPERATURE,h*100,h IS AN INTEGER,DEGREES C? 800
MOISTURE CONTENT FEED,% BY WT? 3.0
MOISTURE CONTENT PRODUCT,% BY WT? .1
SUPERFICIAL VELOCITY GAS OUT,M/S? 1.25
SOLID SPECIFIC HEAT,KJ/(KG*K)? .87
MJ/METRIC T:STEAM,OIL;MJ/NM3:GAS? 32
U.S. $/METRIC T:STEAM,OIL;U.S. $/NM3:GAS? .1
POWER COST,U.S. $/KWH? .06
```

DRYING AIR TEMPERATURE,DEGREES C	100
EXHAUST GAS TEMPERATURE,DEGREES C	70
EVAPORATION,KG/H	598
EFFICIENCY,KJ/KG EVAPORATED	13810
DIAMETER,M	5.30
LENGTH,M	37.10
MEDIUM COST,U.S. $/METRIC T OF PRODUCT	1.29
POWER COST,U.S. $/METRIC T OF PRODUCT	1.53
RELEVANT VARIABLE COSTS,U.S. $/METRIC T OF PRODUCT	2.82
INLET AIR FLOW AT 10 DEGREES C,M3/H	57000
EXHAUST GAS FLOW AT TAOUT,M3/H	82000

DRYING AIR TEMPERATURE,DEGREES C	200
EXHAUST GAS TEMPERATURE,DEGREES C	75
EVAPORATION,KG/H	598
EFFICIENCY,KJ/KG EVAPORATED	7453
DIAMETER,M	2.80

```
LENGTH,M                                              19.60

MEDIUM COST,U.S. $/METRIC T OF PRODUCT                 0.70

POWER COST,U.S. $/METRIC T OF PRODUCT                  0.34

RELEVANT VARIABLE COSTS,U.S. $/METRIC T OF PRODUCT     1.03

INLET AIR FLOW AT 10 DEGREES C,M3/H                   15000

EXHAUST GAS FLOW AT TAOUT,M3/H                        22000




DRYING AIR TEMPERATURE,DEGREES C                       300
EXHAUST GAS TEMPERATURE,DEGREES C                       80
EVAPORATION,KG/H                                       598

EFFICIENCY,KJ/KG EVAPORATED                           6794

DIAMETER,M                                            2.20
LENGTH,M                                             15.40

MEDIUM COST,U.S. $/METRIC T OF PRODUCT                 0.63

POWER COST,U.S. $/METRIC T OF PRODUCT                  0.21

RELEVANT VARIABLE COSTS,U.S. $/METRIC T OF PRODUCT     0.84

INLET AIR FLOW AT 10 DEGREES C,M3/H                   9000

EXHAUST GAS FLOW AT TAOUT,M3/H                        14000




DRYING AIR TEMPERATURE,DEGREES C                       400
EXHAUST GAS TEMPERATURE,DEGREES C                       85
EVAPORATION,KG/H                                       598

EFFICIENCY,KJ/KG EVAPORATED                           6686

DIAMETER,M                                            1.90
LENGTH,M                                             13.30

MEDIUM COST,U.S. $/METRIC T OF PRODUCT                 0.62

POWER COST,U.S. $/METRIC T OF PRODUCT                  0.16

RELEVANT VARIABLE COSTS,U.S. $/METRIC T OF PRODUCT     0.79

INLET AIR FLOW AT 10 DEGREES C,M3/H                   7000

EXHAUST GAS FLOW AT TAOUT,M3/H                        11000




DRYING AIR TEMPERATURE,DEGREES C                       500
EXHAUST GAS TEMPERATURE,DEGREES C                       90
EVAPORATION,KG/H                                       598
```

```
EFFICIENCY,KJ/KG EVAPORATED                            6746

DIAMETER,M                                             1.70
LENGTH,M                                              11.90

MEDIUM COST,U.S. $/METRIC T OF PRODUCT                 0.63

POWER COST,U.S. $/METRIC T OF PRODUCT                  0.14

RELEVANT VARIABLE COSTS,U.S. $/METRIC T OF PRODUCT     0.77

INLET AIR FLOW AT 10 DEGREES C,M3/H                    6000

EXHAUST GAS FLOW AT TAOUT,M3/H                         9000



DRYING AIR TEMPERATURE,DEGREES C                        600
EXHAUST GAS TEMPERATURE,DEGREES C                        95
EVAPORATION,KG/H                                        598

EFFICIENCY,KJ/KG EVAPORATED                            6880

DIAMETER,M                                             1.60
LENGTH,M                                              11.20

MEDIUM COST,U.S. $/METRIC T OF PRODUCT                 0.64

POWER COST,U.S. $/METRIC T OF PRODUCT                  0.13

RELEVANT VARIABLE COSTS,U.S. $/METRIC T OF PRODUCT     0.77

INLET AIR FLOW AT 10 DEGREES C,M3/H                    5000

EXHAUST GAS FLOW AT TAOUT,M3/H                         8000



DRYING AIR TEMPERATURE,DEGREES C                        700
EXHAUST GAS TEMPERATURE,DEGREES C                       100
EVAPORATION,KG/H                                        598

EFFICIENCY,KJ/KG EVAPORATED                            7052

DIAMETER,M                                             1.60
LENGTH,M                                              11.20

MEDIUM COST,U.S. $/METRIC T OF PRODUCT                 0.66

POWER COST,U.S. $/METRIC T OF PRODUCT                  0.12

RELEVANT VARIABLE COSTS,U.S. $/METRIC T OF PRODUCT     0.78

INLET AIR FLOW AT 10 DEGREES C,M3/H                    4000

EXHAUST GAS FLOW AT TAOUT,M3/H                         7000
```

DRYING AIR TEMPERATURE,DEGREES C	800
EXHAUST GAS TEMPERATURE,DEGREES C	105
EVAPORATION,KG/H	598
EFFICIENCY,KJ/KG EVAPORATED	7248
DIAMETER,M	1.50
LENGTH,M	10.50
MEDIUM COST,U.S. $/METRIC T OF PRODUCT	0.68
POWER COST,U.S. $/METRIC T OF PRODUCT	0.11
RELEVANT VARIABLE COSTS,U.S. $/METRIC T OF PRODUCT	0.79
INLET AIR FLOW AT 10 DEGREES C,M3/H	4000
EXHAUST GAS FLOW AT TAOUT,M3/H	7000

References

American Institute of Chemical Engineers (AICHE) (1985). *Equipment Testing Procedure: Continuous Direct-Heat Rotary Dryers,* American Institute of Chemical Engineers, New York.

Buell (1985). Private communication.

Friedman, S. J. and W. R. Marshall, Jr. (1949). Studies in rotary drying. Part I— Holdup and dusting, *Chem. Eng. Prog.,* 45, 482.

Friedman, S. J. and W. R. Marshall, Jr. (1949). Studies in rotary drying. Part II— Heat and mass transfer, *Chem. Eng. Prog.,* 45, 573.

Horgan, T. J. (1928). Rotary dryers, *Trans. Institution of Chem. Engineers* (London), 6, 131.

The Institution of Chemical Engineers / The Association of Cost Engineers (1977). *A New Guide to Capital Cost Estimating,* The Institution of Chemical Engineers, Rugby, England.

McCormick, P.Y. (1962). Gas velocity effects on heat transfer in direct heat rotary dryers, *Chem. Eng. Prog.,* 58, 57.

Miller, C. O., B. A. Smith, and W. H. Schuette (1924). Factors influencing the operation of rotary dryers, *Trans. Am. Institute Chem. Engineers,* 38, 841.

Nonhebel, G. and A. H. H. Moss (1971). *Drying of Solids in the Chemical Industry,* Butterworth, London, England.

Perry, R. H., D. W. Green, and J. O. Maloney (1984). *Perry's Chemical Engineers' Handbook,* McGraw-Hill, New York.

Porter, S. J. (1963). The design of rotary driers and coolers, *Trans. Institution of Chemical Engineers,* 41, 272.

Seaman, W. C. and T. R. Mitchell, Jr. (1954). Analysis of rotary dryer and cooler performance, *Chem. Eng. Prog.,* 50, 467.

Swenson (1985, 1986). Private communications.

Williams-Gardner, A. (1971). *Industrial Drying,* Leonard Hill, London.

7

FLASH DRYING

7.1 Introduction

A general description of flash drying is in Section 7.2 and Section 7.3 contains information on short-time convection drying (several sec). Section 7.4 provides a worked example and Section 7.5 discusses the development of a computer program that is incorporated in Appendix 7.1. Section 7.6 covers investment aspects.

7.2 General description

Basically, a flash dryer processes a continuous flow of particulate material that is dried by contact with warm or hot air while being transported by the airstream (see Fig. 7.1 and Table 7.1). Generally, cooling and drying cannot be combined in one dryer. Consequently, the dried product is often cooled in a separate pneumatic transport system; however, other types of coolers are available. Flash-drying equipment is simple, occupying little space and having few moving parts. It is often feasible to combine both drying and vertical transport. Typical gas exit velocities are in the range 10–30 m/sec. Hence, the flash-drying gas velocities exceed those in rotary and in fluid-bed dryers by a factor of approximately ten. The relatively high gas velocities often require large cyclones or bag houses.

The maximum particle size that can be dried is 1–2 mm because larger particles are not entrained by air and also require a longer drying time than that achieved with flash dryers (see Section 7.3). Unlike rotary dryers,

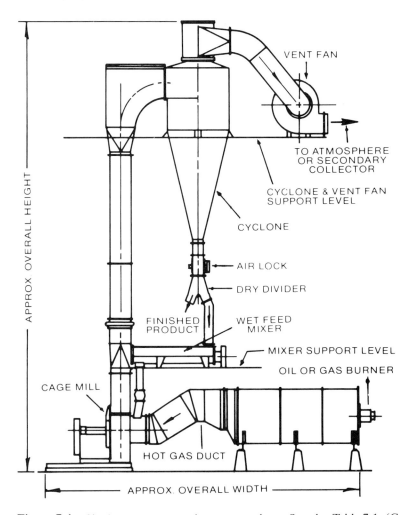

Figure 7.1 Single-stage pneumatic-conveyor dryer. See also Table 7.1. (Courtesy of C-E Raymond, Combustion Engineering, Inc., Chicago, Illinois.)

the flash dryer is susceptible to overloading since the material cannot be transported at high feed rates. Consequently, flash dryers require more attention than rotary dryers.

The high gas velocity can result in abrasion or dust formation, which, in turn, leads to increased maintenance costs. This factor must be considered since it is often stated that maintenance costs are lower for flash dryers than for rotary dryers. Although this finding is, on the surface, correct (because

Table 7.1 Flash Dryer Prices

Cyclone size	Airflow, m³/hr vent	Approximate heat input, MJ/hr	Approximate power requirement, kW	Maximum H₂O evaporation, kg/hr	Approximate overall height, m	Approximate overall width, m	Approximate weight[a], kg	Price CS FOB Shops, $	Price SS FOB Shops, $
3	2,890	1,150	12	320	9.30	7.39	5,450	147,000	157,000
4	5,100	1,970	18	545	10.29	8.26	7,720	160,000	173,000
5	7,990	3,120	28	865	11.05	8.64	9,535	170,000	185,000
6	11,550	4,440	38	1,225	12.04	9.37	12,710	187,000	205,000
7	15,630	6,080	52	1,680	12.73	9.83	14,530	198,000	218,000
8	20,390	7,890	66	2,180	14.33	10.34	16,800	206,000	231,000
9	25,490	9,860	81	2,725	15.62	11.18	19,980	220,000	250,000
10	32,280	11,690	99	3,405	15.93	11.48	23,155	235,000	270,000
12	45,870	18,070	143	4,995	17.83	12.29	29,055	283,000	333,000
14	62,860	24,650	198	6,810	19.91	13.21	33,600	313,000	373,000
16	81,550	31,220	237	8,625	22.35	13.87	39,500	362,000	432,000
18	103,640	39,430	294	10,895	25.30	15.09	52,665	428,000	513,000
20	127,430	49,290	367	13,620	28.19	16.36	56,750	448,000	549,000
22	152,910	59,150	432	16,345	31.24	17.63	61,290		

[a]Weight of cage mill flash-drying system less motor.

Note: Based upon 650°C air-inlet temperatures. For other air-inlet temperatures, water evaporation is proportional to air temperature drop. Delivery comprises direct air heater, cage mill for disintegration of feed, dryer, cyclone, fan, and ducts. Motors and secondary dust collectors supplied by others.

Source: Courtesy of C-E Raymond, Combustion Engineering, Inc., Chicago, Illinois.

rotary dryers do have an important rotating element), it is not always borne out in practice.

Dust formation is promoted by high gas velocity when the smaller particles are eroded from the mother particles. The short residence time is another source of dust formation. This applies when the moisture is a solution of the solid in the liquid and the evaporation proceeds at such a rate that the solid crystallizing from the liquid does not attach itself to the mother crystals but instead forms nuclei. Not only is dust formation a nuisance, it can also be a hazard. However, an attractive aspect concerning process safety is the minimal hold-up of a flash dryer.

The composition of the feed must not cause sticking in the feed section. If the feed is not entirely suitable it can be modified by either backmixing it with the product or using a cage mill to disintegrate it while the drying gas is passed through the mill (approximate maximum air temperature 750°C). Furthermore, these two options can be combined.

The flash-drying system may contain a classifying device that acts on the dried product. When the coarser particles are returned to the feed section to undergo the additional drying often required for larger particles, they may be milled in the return line. A further option is the installation of a slinger below the feed point so that falling particles are disintegrated.

It is also possible to arrange several flash dryers in series to provide additional residence time or to dry at different conditions. In addition, it is also possible to arrange in series a flash dryer and a dryer providing longer residence time, e.g., a fluid-bed dryer (this system is used in some PVC installations).

Usually the air used for drying is sufficient for conveying; however, it is good practice not to exceed a solids/air mass-ratio of one. There are some exceptions to this generalization, e.g., in the drying of relatively coarse material (500 μm particles) containing little moisture (e.g., 2–3 % water by weight) at elevated temperature (ca. 350°C). Kröll (1978) has provided further information on this subject.

The thermal degradation of organic materials is dependent on three factors: (a) time, (b) temperature, and (c) concentration. A high air-inlet temperature can sometimes be chosen because most of the evaporation occurs while the particles are at the wet-bulb temperature. Another reason, specific to flash dryers, is the short drying time. The air-outlet temperature usually exceeds that of the product by 10–30°C.

There is usually a slight negative pressure at the feed point for two reasons: first, the underpressure prevents incrustations in the feed section; and second, it prevents dust emissions at the feed point. The drying system may contain only one fan, typically located between the cyclones and the

Figure 7.2   Flash dryer installation for the drying of potato starch. (Courtesy of Bépex Corporation, Minneapolis, Minnesota.)

wet scrubber. Alternatively, the introduction of both primary and secondary air into the combustion chamber may require additional fans.

The diameter of the drying tube can reach 1 m and the length can vary between 10 and 30 m (see Fig. 7.2 which illustrates a conventional flash-dryer installation). An effective way to control the drying operation is via the fuel or the steam flow on the basis of the air-outlet temperature. The dryer must be able to cope with varying feed flows because there is usually no buffer between the liquid/solid separating system and the dryer; hold-up of wet solids cannot be easily controlled. It is a good practice not to vary the airflow because this affects the transport function. Furthermore, it is important that the dryer be internally smooth to avoid incrustations. Lastly, for maintenance purposes, flash dryers may require up to 5% of the investment annually.

7.3   Drying in seconds

In this section we will consider the flash drying of an organic material consisting of spherical particles with a diameter of 250 μm and having an inlet moisture content of 10% by weight. The air-inlet temperature is 225°C; the heat transfer is rate-determining.

Basic data

Solid

Insoluble in water

250–μm spheres

ρ_s = 1,200 kg/m^3

λ_s = 0.2 W/m · K

CP = 1.2 kJ/kg · K

Air

R_A = 1.29 kg/m^3 at 0°C and 1 bara

C_p = 1.0 kJ/kg · K 0–200°C

η_a = 21.8 × 10^{-6} N · sec/m^2 at 100°C
      25.9 × 10^{-6} N · sec/m^2 at 200°C

λ_a = $\left.\begin{array}{l} 0.0300 \\ 0.0338 \\ 0.0373 \\ 0.0407 \end{array}\right\}$ W/m · K at $\left\{\begin{array}{l} 77°C \\ 127°C \\ 177°C \\ 227°C \end{array}\right.$

Water

CP_v = 1.886 kJ/kg · K

CP_w = 4.19 kJ/kg · K

ΔH = 2,504 kJ/kg at 0°C

Operational data

Feed

20°C

A_1 = 10% by wt

Product

2,000 kg/hr

60°C

A_2 = 0.5% by wt

Ambient air

10°C

60% relative humidity

Warm air

Indirect steam heating

225°C

Spent drying gas

77°C

Mass balance, kg/hr

	In	Out
H_2O	221	10
Solids	1,990 +	1,990 +
	2,211	2,000

Water evaporation $221 - 10 = 211$

Net heat, kJ/hr

Net heat is the heat absorbed by the process flow. H_2O warming, evaporation, and vapor warming:

$$211 (2,504 + 1.886 \times 77 - 4.19 \times 20) = 541,304$$
Solid warming:
$$1,990 \times 1.2 (60 - 20) = 95,520$$
Remaining water warming:
$$10 \times 4.19 (60 - 20) = \underline{1,676} +$$
$$638,500$$

Temperature profiles

See Figure 7.3. Adiabatic saturation temperature (wet-bulb temperature) 47°C. See Mollier's diagram.

Heat transfer

$Qtot_1 = U \times A \times (\Delta T)_m$
$Qtot_1 =$ net heat, i.e., 638,500 kJ/hr
$U =$ heat-transfer coefficient; $kJ/m^2 \cdot hr \cdot K$, which is calculated by using Froessling's equation (Froessling, 1938).
$A =$ area available in the dryer for heat transfer, m^2, to be calculated.
$(\Delta T)_m =$ logarithmic mean-temperature difference, calculated by taking the wet-bulb temperature as the solid's temperature (free-moisture evaporation).

$$(\Delta T)_m = \frac{(225 - 47) - (77 - 47)}{\log_e \frac{(225 - 47)}{(77 - 47)}} = 83.1 \text{ K}$$

Notes:
1. Calculating $(\Delta T)_m$ as indicated is a fair approximation (Perry et al., 1984).
2. U is larger in the feed section than in the upper part of the drying tube because of the initial large velocity difference. This effect is neglected, leading to a conservative A value.

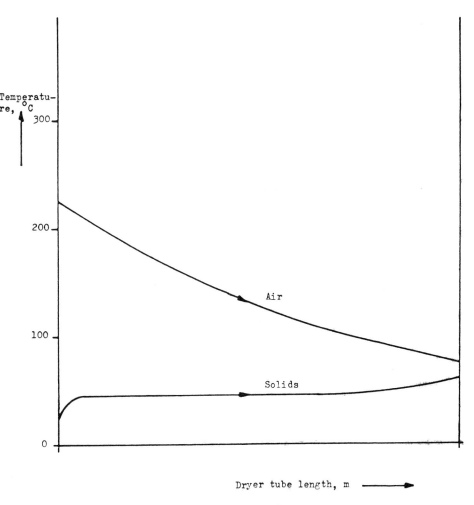

Figure 7.3 Temperature profiles plotted according to the sample calculation in Section 7.3.

Froessling's equation:

$Nu = 2 + 0.552 \times Pr^{1/3} \times Re^{1/2}$. Only dry air is considered.

$$Re = \frac{R_A \times v \times d_{50}}{\eta_a}$$

$$\text{calculation at } \frac{225 + 77}{2} \quad = 151°C$$

$$R_A = 1.29 \times \frac{273}{273 + 151} = 0.83 \text{ kg/m}^3$$

$d_{50} = 250 \times 10^{-6} \text{ m}$

$\eta_a = 23.9 \times 10^{-6} \text{ N} \cdot \text{sec/m}^2$

terminal velocity is approximately 1 m/sec (see Table 4.1)

$$Re = \frac{0.83 \times 1 \times 250 \times 10^{-6}}{23.9 \times 10^{-6}} = 8.68$$

$$Pr = \frac{\eta_a \times C_p}{\lambda_a} = \frac{23.9 \times 10^{-6} \times 1.0 \times 10^3}{0.0356} = 0.671$$

$$Nu = 2 + 0.552 \times 0.671^{1/3} \times 8.68^{1/2} = 3.42$$

$$Nu = \frac{U \times d_{50}}{\lambda_a} \rightarrow U = \frac{Nu \times \lambda_a}{d_{50}} = \frac{3.42 \times 0.0356}{250 \times 10^{-6}} = 487.0 \text{ W/m}^2 \cdot \text{K}$$

$$A_{req} = \frac{Qtot_1}{U \times (\Delta T)_m} = \frac{638,500 \times 10^3}{3,600 \times 487.0 \times 83.1} = 4.38 \text{ m}^2$$

Area passed through the dryer per hr:

$$\frac{1,990 \times \pi (250 \times 10^{-6})^2 \, 6}{\pi \times (250 \times 10^{-6})^3 \times 1,200} = 39,800 \text{ m}^2/\text{hr}$$

Required residence time:

$$\tau_{req} = \frac{4.38 \times 3,600}{39,800} = 0.40 \text{ sec}$$

This example qualitatively exhibits that the drying of small particles in a flash dryer proceeds at a high rate. However, the required drying time rapidly increases with an increase in particle size.

Additional remarks

Drying in a flash dryer must be completed instantaneously, which means that the particles must be small enough to allow a rapid and complete evaporation of the free moisture and that the internal diffusion of moisture is not limiting.

Furthermore, it is often necessary to raise the temperature of the solid to a specific value to be certain of a final moisture content. Again, the particles must be small enough so that the thermal conductivity of the solids does not control during the heating operation. Mass transfer is often the rate-determining step in the final stage of the drying process.

Biot's number provides the possibility of an assessment of the relative importance of the heat transfer to the particle and of the heat dissipation in the particle itself.

$$Bi = \frac{U \times d_{50}}{2 \times \lambda_s}$$

For the case considered: $Bi = \dfrac{487.0 \times 250 \times 10^{-6}}{2 \times 0.2} = 0.30$

If $Bi < 0.1$, the heat transfer to the particle is the rate-determining step. However, when $Bi = 1$, this statement is still approximately correct.

Heat dissipation in the particles considered cannot be neglected entirely, which is typical for organic materials.

Typical conductivities of inorganic materials: (a) salt (sodium chloride) 5.8 W/m · K at 30°C, and (b) sand 1.4 W/m · K at 30°C.

Biot numbers for inorganic materials tend to be much lower than those for organic materials.

Froessling investigated the evaporation of droplets of nitrobenzene, aniline, and water.

Droplet sizes: 0.1–0.9 mm
Air velocities: 0.2–7 m/sec
Re: 2–800

For the evaporation of naphthalene spheres, the same result was obtained, with the temperature of the experiments being 20°C.

7.4 Worked example

Product

General: 5 metric tons/hr of an inorganic material
Temperature as fed: 20°C
Particle size: 50% by wt larger than 600 μm
Initial moisture content: 15% by wt (as fed)
Final moisture content: 0.1% by wt (as discharged)
Specific heat of solids: 0.8 kJ/kg · K
Water solubility of solids:0 % by wt

Process

Ambient temperature: 10°C
Ambient relative humidity: 50%
Inlet gas temperature: 600°C

Water-air data

Water specific heat: 4.19 kJ/kg · K
Water vapor specific heat: 1.886 kJ/kg · K
Latent heat of evaporation at 0°C: 2,504 kJ/kg
Water vapor specific mass relationship at 10^5 Pa: $220/(273 + T)$ kg/m^3
Air specific heat (average): 1.05 kJ/kg · K
Air specific mass relationship at 10^5 Pa: $355/(273 + T)$ kg/m^3

Natural gas data

Calorific value: 32 MJ/nm^3
Specific mass at 0°C and 10^5 Pa: 0.78 kg/m^3

After combustion, 1 nm^3 of gas provides 1.67 nm^3 of water vapor. 1 nm^3 of gas requires 8.50 nm^3 of air (50% relative humidity at 10°C, i.e., 0.004 kg of H_2O/kg of dry air) for stoichiometric combustion.

Pilot-plant data

Exit-gas velocity: 20 m/s
Gas-outlet temperature: 120°C
Product-outlet temperature: 90°C (product is dry)
Material is not thermally sensitive

No recycle

Mass balance, kg/hr

	In	Out
H_2O	881	5
Solids	4,995	4,995
	───── +	───── +
	5,876	5,000

Evaporation $881 - 5 = 876$

Net heat, kJ/hr

$Q_1 = 876(2,504 + 1.886 \times 120 - 4.19 \times 20) = 2,318,352$

$Q_2 = 4,995 \times 0.8(90 - 20) \qquad\qquad = \quad 279,720$

$Q_3 = 5 \times 4.19(90 - 20) \qquad\qquad\quad = \qquad 1,467$

$\qquad\qquad\qquad\qquad\qquad\qquad\qquad\qquad\qquad\qquad +$

$Qtot_1 \qquad\qquad\qquad\qquad\qquad\qquad\qquad 2,599,539$

$$Qtot_2 = \frac{1.25 \times (600 - 10)}{(600 - 120)} \times 2,599,539 = 3,994,083$$

$Qtot_2$ is the heat to be supplied in the combustion chamber by the combustion of natural gas (see Section 7.5 for a discussion of the heat-loss factor 1.25). Natural gas consumption

$$\frac{3,994,083}{32,000} = 124.8 \text{ nm}^3/\text{hr.}$$

Dew point exhaust gas, °C

Stoichiometric natural-gas combustion

	kg/hr	
	In	Out
Natural gas	97.34	
Dry air	1,362.98	
Water vapor	5.44	173.17
Combustion products		1,292.59
	1,465.76 +	1,465.76 +

Total amount of gases from the combustion chamber:

$$\frac{3,994,083}{1.05(600 - 10)} = 6,447.27 \text{ kg/hr.}$$

Amount of secondary air: $6,447.27 - 1,465.76 = 4,981.51$ kg/hr (composed of 19.80 kg/hr of water vapor and 4,961.71 kg/hr of dry air). Air leaks into the dryer amount to 20% of the total amount of gases from the combustion chamber: $0.2 \times 6,447.27 = 1,289.45$ (composed of 5.14 kg/hr of water vapor and 1,284.31 kg/hr of dry air).

Process flow	Dry	H_2O	Total
Combustion	1,292.59	173.17	1,465.76
Secondary air	4,961.71	19.80	4,981.51
Leakage	1,284.31	5.14	1,289.45
Evaporation	0	876.00	876.00
	7,538.61 +	1,074.11 +	8,612.72 +

Water content $\dfrac{1,074.11}{7,538.61} = 0.142$ kg/kg of dry air.

Temperature: 120°C
Dewpoint: 58°C

Sizing drying-gas preparation unit

$Qtot_2$ = 3,994,083 kJ/hr.

Buy a combustion chamber having a capacity of 5,000 MJ/hr or 1,400 kW (25% spare capacity).

Gas flow from the combustion chamber is 6,447.27 kg/hr.

R_A = 355/ (273 + 10) = 1.25 kg/m^3

Volume flow 6,447.27 / 1.25 = 5,157.82 m^3/hr

Fan power $\dfrac{5,157.82 \times 2,500}{3,600 \times 1,000 \times 0.5}$ = 7.2 kW

Take a fan with a motor of 10 kW.

Dryer sizing

Upward gas flow including leakage (excluding evaporation):

1.2 × 6,447.27 = 7,736.72 kg/hr

R_A = 355/ (273 + 120) = 0.903 kg/m^3

Volumetric flow 7,736.72/0.903 = 8,567.80 m^3/hr

Upward evaporated water flow:

876 kg/hr

R_W = 220/ (273 + 120) = 0.560 kg/m^3

Volumetric flow 876/0.560 = 1,564.29 m^3/hr

Total upward gas flow:

8,567.80 + 1,564.29 = 10,132.09 m^3/hr

$\pi/4 \times D^2 \times 20 \times 3600$ = 10,132.09 → D = 0.42 m

Take D = 0.40 m

Pilot-plant dryer height is 10 m. Take industrial dryer-height of 12 m.

Sizing exhaust gas unit

Fan power: $\dfrac{10,132.09 \times 3,000}{3,600 \times 1,000 \times 0.5}$ = 16.9 kW

Take a fan with a motor of 25 kW.

Choose a cyclone.

Survey

Dryer diameter: 0.4 m

Drying-tube length: 12 m

Capacity of combustion chamber: 1,400 kW

Natural gas consumption (load factor 1.5):

$\dfrac{1.5 \times 124.8}{5}$ = 37.4 nm^3/ton of product

For $0.1 /nm^3 of gas: 3.74 $/t

Electricity consumption: $1.5(7.2 + 16.9 + 10.0) = 51.2$ kW (including 10.0 kW miscellaneous).

For $ 0.06/kWhr: $\dfrac{51.2 \times 0.06}{5} = 0.61$ $/t

After examining Table 7.1, we find that a stainless steel dryer costing $205,000, (FOB) is adequate. With installation costs: $2.25 \times 205,000 = \$461,250$.

The fixed costs for an annual production of $7,000 \times 5 = 35,000$ tons:

$$\frac{0.35 \times 461,250}{35,000} = \$4.61$$

Recycle

Introduction

During an experiment, it was found that recycling about 50% of the exhaust gases is possible. The exhaust-gas temperature had to be increased to 125°C to attain, in the product, 0.1% H_2O by wt. The product exit temperature is 95°C; the dewpoint of the exhaust gas is 70°C.

Mass balance, kg/hr

	In	Out
H_2O	881	5
Solids	4,995	4,995
	+	+
	5,876	5,000

Evaporation: $881 - 5 = 876$.

Net heat, kJ/hr

$Q_1 = 876(2,504 + 1,886 \times 125 - 4.19 \times 20)$ $=$ 2,326,612
$Q_2 = 4,995 \times 0.8(95 - 20)$ $=$ 299,700
$Q_3 = 5 \times 4.19(95 - 20)$ $=$ 1,571 $+$

Total   2,627,883

At this stage, the assumption is that 60% of the net heat is supplied by the recycled and reheated exhaust gas. The other 40% is supplied by gases starting at a temperature of 10°C. Heat input in the combustion chamber to the recycled gas is $1.25 \times 0.6 \times 2,627,883 = 1,970,912$ kJ/hr.

Heat input in the combustion chamber to the fresh gases is

$$1.25 \times 0.4 \times \frac{(600 - 10)}{(600 - 125)} \times 2,627,883 = 1,632,054 \text{ kJ/hr}$$

This adds up to:

1,970,912 + 1,632,054 = 3,602,966 kJ/hr.
Natural gas consumption is

$$\frac{3,602,966}{32,000} = 112.6 \text{ nm}^3/\text{hr.}$$

Per metric ton of product in the long run: $\dfrac{1.5 \times 112.6}{5} = 33.8 \text{ nm}^3.$

By recycling, annual savings of 0.1(37.4 − 33.8)35,000 = $12,600. To gain an advantage, an extra investment of $25,000 to $40,000 is possible.

7.5   Computer program

The computer program in BASIC is provided in Appendix 7.1. Reference is made to the description of the program for a continuous rectangular fluid-bed dryer (see Section 5.6; Section 4.2 may also be consulted).

Additional remarks.

Input: lines 60–160.
Mass-balance process flow: 180–220.
Line 270 relates the drying-gas exit temperature to the drying-gas inlet temperature: $TA_{OUT} = 0.1875 \times TA_{IN} + 35$.

Figure 7.4 plots this relationship, with the measured points taken from the literature (Williams-Gardner, 1971; Nonhebel and Moss, 1971; Noden, 1972, 1974; Meedom, 1977) and recently observed for industrial dryers.

As can be seen in Figure 7.4, there is a considerable spread in the measured points. The corresponding data for rotary dryers and fluid-bed dryers are more homogeneous. The given relationship is an approximation of the actual state of affairs, with the explanation given that TA_{OUT} for flash dryers is often chosen to guarantee a desired product-moisture content; a large exit driving-force in the dryer is often mandatory to obtain the desired process result in a short time.

Of course, there is a direct relationship between the air- and the product-outlet temperatures. If a specific flash dryer has air- and product-outlet temperatures (20–30°C) considerably deviating from the predictions based on the relationship $TA_{OUT} = 0.1875 \times TA_{IN} + 35$, and if the feed-moisture content is low (2–4% by wt), the impact on the net heat (process-flow heat requirement) is considerable. This situation arises because, in these instances, the heat effect that is associated with the heating up of the solids is important in relationship to the evaporation heat-requirement. Examples include the flash drying of vacuum pan salt and anhydrous sodium sulphate. Data reported in the literature (17 points) and measured in Akzo processing

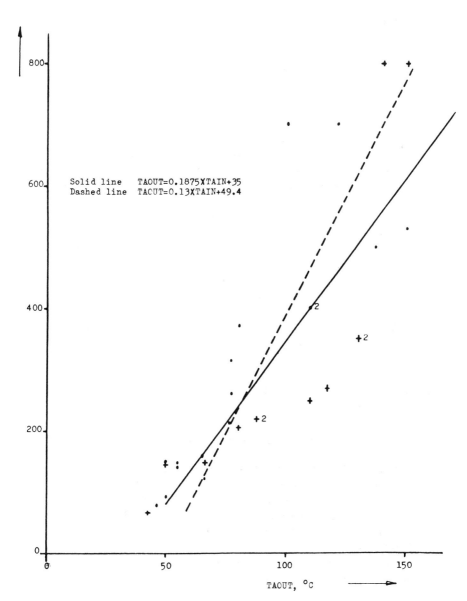

Figure 7.4   Relationship between air inlet and air outlet temperatures for flash dryers. Solid line: $TA_{OUT} = 0.1875 \times TA_{IN} + 35$; dashed line: $TA_{OUT} = 0.13 \times TA_{IN} + 49.4$.

plants (12 points) were analyzed statistically. The relationship $TA_{OUT} = 0.13 \times TA_{IN} + 49.4$ holds (linear regression), and the correlation coefficient is 0.82. Instead of this relationship, the equation $TA_{OUT} = 0.1875 \times TA_{IN} + 35$ is applied. This approach is adopted in order to be relatively safe.

Comparison of the two relationships

TA_{IN}, °C	200	600
$TA_{OUT} = 0.1875 \times TA_{IN} + 35$	72.5	147.5
$TA_{OUT} = 0.13 \times TA_{IN} + 49.4$	75.4	127.4

The restricted applicability (see Fig. 7.4) of the relation at low-temperature drying as pointed out in Section 5.6 must be mentioned as well. Lines 280 through 310 concern the process-stream enthalpy change. The specific item for the flash dryer is the assumption that the product's temperature has a backlog of 30°C as compared to the air-outlet temperature, which is about right, with somewhat lower values found for equipment with air-inlet temperatures of 100–200°C although somewhat higher figures can be found with air-inlet temperatures of 600–800°C. Line 320 calculates the total amount of heat to be put into the drying-gas flow. The remarks regarding the choice of 10°C as the ambient temperature in Section 6.5 apply here as well. The factor of 1.2 allows for the steady-state heat losses, which can be calculated as an average value from measurements reported in the literature (Williams-Gardner, 1971; Noden, 1972, 1974). The implicit assumption is that the reported primary energy consumption of the 17 dryers was measured over a relatively short period and does not reflect the long-term consumption.
    Assumptions

1. Ambient temperature is 10°C.
2. $Tf = 20°C$.
3. $TP_{OUT} = TA_{OUT} - 30$.
4. Reported TA_{IN}- and TA_{OUT}-values were taken.

The arithmetic average of the ratio calculated consumption to the indicated consumption is 1.0, with the standard deviation being 0.10. However, the aforementioned factor of 1.2 is an average. Higher values apply for relatively small evaporation loads (e.g., 1.4 for 100–200 kg of water evaporation/hr). For large evaporation loads (e.g., 4–6 tons/hr), a value of 1.1 is often found.
    Lines 340 through 420 concern the calculation of the diameter of the drying tube. The airflow is multiplied by 1.2 to allow for the attraction of ingress air.
    In line 380, the water vapor flow and the airflow are added up. Lines 440

U.S.\$/t of salt

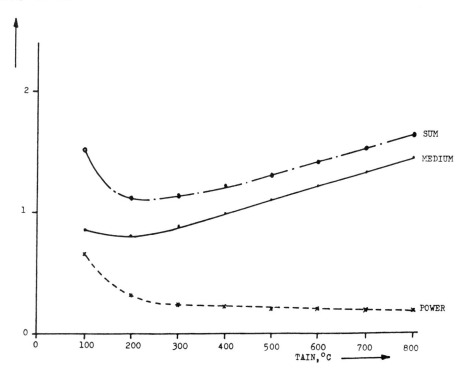

Figure 7.5 Flash-drying costs. See also Appendix 7.1.

and 450 concern the heating-medium cost, multiplied with *Lofa* (the load factor) to allow for the increasing medium cost per metric ton of the product (see background explanation in Section 4.2). The load factor is introduced into the program in line 70. One must choose between 1.25, 1.5, and 1.75. This advice is based on an analysis of the long-term primary-energy consumption of 7 flash dryers in the process industry.

The next item to be calculated is the power cost, and line 480 gives the total flash-dryer power consumption. This formula consists of the power consumption of the main (exhaust) fan multiplied by 2 to make an allowance for the fans to convey the air to the burner, a cage mill, feeders, screw conveyors, and the like. A dryer pressure-drop of 300 mm wg is used. This formula was proposed after checking the reported sizes of fan motors (Williams-Gardner, 1971), the reported total dryer secondary-energy consumptions (Noden, 1972, 1974), and the installed power in Akzo flash-drying plants. Line 490 indicates the power cost in dollars/metric ton of product. Finally, the two variable cost factors are added in line 510. Line

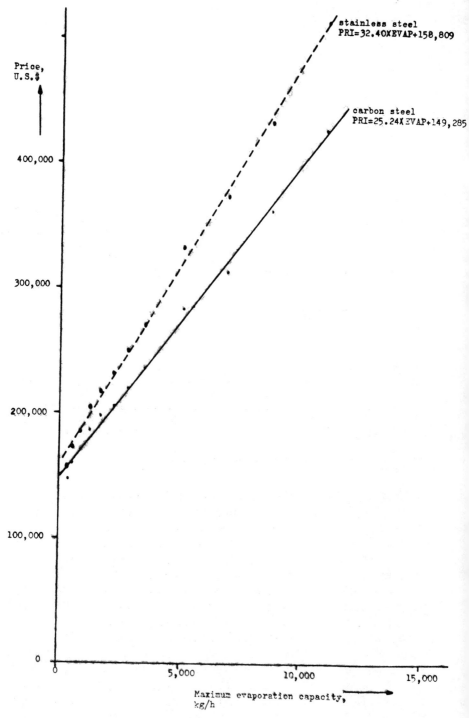

Figure 7.6 Relationship between evaporation capacity and price for flash dryers. See also Table 7.1. (Courtesy of C-E Raymond, Combustion Engineering, Inc., Chicago, Illinois.)

520 gives the weight ratio of the dried product flow and the exit-gas flow. Powder transport requires a number not exceeding 1 (Kröll, 1978). Lastly, Figure 7.5 plots the variable costs as a function of the air-inlet temperature for the example in Appendix 7.1.

7.6 Investment aspects

Table 7.1 provides prices for both carbon steel and stainless steel constructions (C-E Raymond, Combustion Engineering, Inc., 1986). The table is based on 650°C air-inlet temperature. The maximum water evaporation in kg/hr at other air-inlet temperatures can be calculated by assuming that the maximum water evaporation is (approximately) proportional to the air temperature drop. For other air-inlet temperatures, the equivalent maximum water evaporation can be calculated:

$$Syn_{EVAP} = (Evap \times 493.1) / (0.8125 \times TA_{IN} - 35) \text{ kg/hr}$$

The relationship $TA_{OUT} = 0.1875 \times TA_{IN} + 35$ was used to arrive at the expression for Syn_{EVAP}.

Table 7.1 relates to maximum water evaporations only, i.e., the heating of solids is neglected. As pointed out earlier, this aspect can play an important role in the drying of materials containing little moisture (e.g., vacuum pan salt).

Figure 7.6 contains a plot of the data taken from Table 7.1. The relationships between the prices of both carbon steel and of stainless steel dryers and the maximum water evaporation appear to be approximately linear. The numeric expressions were obtained by means of linear regression, with the correlation coefficients, in both cases, being 0.998. The figures for the size-20 cyclone were not considered because they were not satisfactory vis à vis other data.

Notation

A	Particle heat-transfer area	m²
A_{req}	Required heat-transfer area	m²
A_1	Feed moisture content	% by wt
A_2	Product moisture content	% by wt
$Airflo_2$	Airflow leaving the dryer	m³/hr
Bi	Biot number	
Cal_{VAL}	Fuel or steam calorific value	mJ/t or mJ/nm³
Cap	Product-mass flow	kg/hr
CP	Solid specific heat	kJ/kg · K
CP_v	Water-vapor specific heat	kJ/kg · K

CP_w	Water specific heat	kJ/kg · K
C_p	Air specific heat	kJ/kg · K
D	Dryer diameter	m
D_1	Dryer diameter	m
D_2	Dryer diameter	
D_3	Dryer diameter	
$Diam$	Dryer diameter	m
d_{50}	Average particle size (weight basis)	μm
$Effic$	Dryer efficiency	kJ/kg
$Evap$	Dryer water-evaporation load	kg/hr
$Gasflo_2$	Gas flow leaving the dryer	m^3/hr
h	Integer in FOR-NEXT loop	
$Lofa$	Load factor	
$Medco$	Medium cost	\$/nm^3 or \$/t
$Medcos$	Medium cost per metric ton of product	\$/t
$Medflo$	Medium flow	nm^3/hr or t/hr
Nu	Nusselt number	
Pow	Power consumption	kW
Pow_{CO}	Power cost	\$/kWhr
Pow_{COS}	Power cost per metric ton of product	\$/t
Pr	Prandtl number	
PR_I	Price	\$
Q_1	Heat flow to the evaporated water	kJ/hr
Q_2	Heat flow to the dry solids	kJ/hr
Q_3	Heat flow to the residual water	kJ/hr
$Qtot_1$	Net heat	kJ/hr
$Qtot_2$	Steady-state dryer-heat requirement	kJ/hr
R_A	Air specific mass	kg/m^3
R_W	Water vapor specific mass	kg/m^3
Re	Reynolds number	
Sol	Dry-solids flow	kg/hr
Syn_{EVAP}	Recalculated water evaporation	kg/hr
T	Temperature	°C
TA_{IN}	Drying-air temperature	°C
TA_{OUT}	Spent drying-gas temperature	°C
Tf	Feed temperature	°C
T_{MAX}	Maximum drying-air temperature	°C
TP_{OUT}	Product exit temperature	°C
U	Heat-transfer coefficient	kJ/m^2 · hr · K or W/m^2 · K
	Superficial air velocity out	m/sec
v	Terminal velocity of a falling particle	m/sec

Var_{COS}	Relevant variable costs per metric ton of product	[US]$/t
WA_{FLO}	Evaporated water flow	m³/hr
W_{IN}	Water flow to the dryer	kg/hr
W_{OUT}	Water flow leaving the dryer	kg/hr
ΔH	Water latent heat of evaporation at 0°C	kJ/kg
$(\Delta T)_m$	Logarithmic mean temperature difference	K
λ_a	Air thermal conductivity	W/m · K
λ_s	Solid thermal conductivity	W/m · K
η_a	Air viscosity	N · sec/m²
ρ_s	Solid specific mass	kg/m³
τ_{req}	Required residence time	sec

Appendix   7.1

```
0
LIST
10   WIDTH "LPT1:",132
20   LPRINT CHR$(15);
30   LPRINT "FLASH DRYER SCREENING PROGRAM"
40   LPRINT
50   LPRINT
60   INPUT "KG/H OUT";CAP
70   INPUT "LOAD FACTOR;FULL:1.25,NORMAL:1.5,LOW:1.75";LOFA
80   INPUT "FEED TEMPERATURE,DEGREES C";TF
90   INPUT "MAXIMUM DRYING AIR TEMPERATURE,h*100,h IS AN INTEGER,DEGREES C";TMAX
100  INPUT "MOISTURE CONTENT FEED,% BY WT";A1
110   INPUT "MOISTURE CONTENT PRODUCT,% BY WT";A2
120  INPUT "EXIT DRYING TUBE AIR VELOCITY,M/S";U
130  INPUT "SOLID SPECIFIC HEAT,KJ/(KG*K)";CP
140  INPUT "MJ/METRIC T:STEAM,OIL;MJ/NM3:GAS";CALVAL
150  INPUT "U.S. $/METRIC T:STEAM,OIL;U.S. $/NM3:GAS";MEDCO
160  INPUT "POWER COST,U.S. $/KWH";POWCO
170  LPRINT
180  REM MASS BALANCE IN KG/H
190  WOUT=.01*CAP*A2
200  SOL=CAP-WOUT
210  WIN=SOL/(100-A1)*A1
220  EVAP=WIN-WOUT
230  LPRINT
240  REM STUDY OF THE DRYING AIR TEMPERATURE EFFECT
250  FOR TAIN=100 TO TMAX STEP 100
260    REM HEAT BALANCE IN KJ/H
270    TAOUT=.1875*TAIN+35
280    Q1=EVAP*(2504+1.886*TAOUT-4.19*TF)
290    Q2=SOL*CP*(TAOUT-30-TF)
300    Q3=WOUT*4.19*(TAOUT-30-TF)
310    QTOT1=Q1+Q2+Q3
320    QTOT2=QTOT1*(TAIN-10)/(TAIN-TAOUT)*1.2
330    REM DRYER SIZING IN M
340    RA=355/(TAOUT+273)
350    AIRFLO2=QTOT2/((TAIN-10)*1.05*RA)*1.2
360    RW=220/(TAOUT+273)
```

```
370   WAFLO=EVAP/RW
380   GASFLO2=AIRFLO2+WAFLO
390   D1=SQR(GASFLO2/(U*ATM(1)*3600))
400   D2=D1/.1
410   D3=INT(D2)+1
420   DIAM=.1*D3
430   REM HEATING MEDIUM COST IN U.S. $ PER METRIC T
440   MEDFLO=QTOT2/(CALVAL*1000)
450   MEDCOS=MEDFLO*MEDCO/CAP*1000*LOFA
460   EFFIC=MEDFLO*CALVAL*1000/EVAP*LOFA
470   REM POWER COST IN U.S. $ PER METRIC T
480   POW=2*GASFLO2*3000/(3600*.5*1000)
490   POWCOS=POW*POWCO/CAP*1000*LOFA
500   REM ADDING COSTS IN U.S. $ PER METRIC T
510   VARCOS=MEDCOS+POWCOS
520   RATIO=CAP/(EVAP+AIRFLO2*RA)
530   REM OUTPUT
540   LPRINT
550   LPRINT
560   LPRINT
570   LPRINT USING "DRYING AIR TEMPERATURE,DEGREES C                    ####";TAIN
580   LPRINT USING "EXHAUST GAS TEMPERATURE,DEGREES C                    ###";TAOUT
590   LPRINT USING "EVAPORATION,KG/H                                    ####";EVAP
600   LPRINT
610   LPRINT USING "EFFICIENCY,KJ/KG EVAPORATED                       #####";EFFIC
620   LPRINT
630   LPRINT USING "DIAMETER,M                                        #.##";DIAM
640   LPRINT
650   LPRINT USING "MEDIUM COST,U.S. $/METRIC T OF PRODUCT          ###.##";MEDCOS
660   LPRINT
670   LPRINT USING "POWER COST,U.S. $/METRIC T OF PRODUCT           ##.##";POWCOS
680   LPRINT
690   LPRINT
700   LPRINT USING "RELEVANT VARIABLE COSTS,U.S. $/METRIC T OF PRODUCT  ###.##";VARCOS
710   LPRINT
720   LPRINT USING "EXHAUST GAS FLOW AT TAOUT,M3/H                  ######";GASFLO2
730   LPRINT
740   LPRINT USING "WEIGHT RATIO SOLIDS/GAS                           ##.#";RATIO
750   NEXT TAIN
760   LPRINT
770   WIDTH "LPT1:",80
780   LPRINT CHR$(18)
790 END
0
```

```
RUN
FLASH DRYER SCREENING PROGRAM


KG/H OUT? 2000
LOAD FACTOR;FULL:1.25,NORMAL:1.5,LOW:1.75? 1.5
FEED TEMPERATURE,DEGREES C? 20
MAXIMUM DRYING AIR TEMPERATURE,h*100,h IS AN INTEGER,DEGREES C? 800
MOISTURE CONTENT FEED,% BY WT? 3.0
MOISTURE CONTENT PRODUCT,% BY WT? .1
EXIT DRYING TUBE AIR VELOCITY,M/S? 20
SOLID SPECIFIC HEAT,KJ/(KG*K)? .87
```

```
MJ/METRIC T:STEAM,OIL;MJ/NM3:GAS? 32
U.S. $/METRIC T:STEAM,OIL;U.S. $/NM3:GAS? .1
POWER COST,U.S. $/KWH? .06
```

DRYING AIR TEMPERATURE,DEGREES C	100
EXHAUST GAS TEMPERATURE,DEGREES C	54
EVAPORATION,KG/H	60
EFFICIENCY,KJ/KG EVAPORATED	9216
DIAMETER,M	0.30
MEDIUM COST,U.S. $/METRIC T OF PRODUCT	0.86
POWER COST,U.S. $/METRIC T OF PRODUCT	0.66
RELEVANT VARIABLE COSTS,U.S. $/METRIC T OF PRODUCT	1.52
EXHAUST GAS FLOW AT TAOUT,M3/H	4383
WEIGHT RATIO SOLIDS/GAS	0.4

DRYING AIR TEMPERATURE,DEGREES C	200
EXHAUST GAS TEMPERATURE,DEGREES C	73
EVAPORATION,KG/H	60
EFFICIENCY,KJ/KG EVAPORATED	8622
DIAMETER,M	0.20
MEDIUM COST,U.S. $/METRIC T OF PRODUCT	0.81
POWER COST,U.S. $/METRIC T OF PRODUCT	0.32
RELEVANT VARIABLE COSTS,U.S. $/METRIC T OF PRODUCT	1.12
EXHAUST GAS FLOW AT TAOUT,M3/H	2106
WEIGHT RATIO SOLIDS/GAS	0.9

DRYING AIR TEMPERATURE,DEGREES C	300
EXHAUST GAS TEMPERATURE,DEGREES C	91
EVAPORATION,KG/H	60
EFFICIENCY,KJ/KG EVAPORATED	9495
DIAMETER,M	0.20

```
MEDIUM COST,U.S. $/METRIC T OF PRODUCT            0.89

POWER COST,U.S. $/METRIC T OF PRODUCT             0.24


RELEVANT VARIABLE COSTS,U.S. $/METRIC T OF PRODUCT   1.13

EXHAUST GAS FLOW AT TAOUT,M3/H                    1630

WEIGHT RATIO SOLIDS/GAS                            1.3



DRYING AIR TEMPERATURE,DEGREES C                   400
EXHAUST GAS TEMPERATURE,DEGREES C                  110
EVAPORATION,KG/H                                    60

EFFICIENCY,KJ/KG EVAPORATED                       10603

DIAMETER,M                                         0.20

MEDIUM COST,U.S. $/METRIC T OF PRODUCT            0.99

POWER COST,U.S. $/METRIC T OF PRODUCT             0.22

RELEVANT VARIABLE COSTS,U.S. $/METRIC T OF PRODUCT   1.21

EXHAUST GAS FLOW AT TAOUT,M3/H                    1440

WEIGHT RATIO SOLIDS/GAS                            1.5



DRYING AIR TEMPERATURE,DEGREES C                   500
EXHAUST GAS TEMPERATURE,DEGREES C                  129
EVAPORATION,KG/H                                    60

EFFICIENCY,KJ/KG EVAPORATED                       11792

DIAMETER,M                                         0.20

MEDIUM COST,U.S. $/METRIC T OF PRODUCT            1.10

POWER COST,U.S. $/METRIC T OF PRODUCT             0.20


RELEVANT VARIABLE COSTS,U.S. $/METRIC T OF PRODUCT   1.30

EXHAUST GAS FLOW AT TAOUT,M3/H                    1350

WEIGHT RATIO SOLIDS/GAS                            1.7



DRYING AIR TEMPERATURE,DEGREES C                   600
EXHAUST GAS TEMPERATURE,DEGREES C                  148
EVAPORATION,KG/H                                    60
```

```
EFFICIENCY,KJ/KG EVAPORATED                          13017

DIAMETER,M                                            0.20

MEDIUM COST,U.S. $/METRIC T OF PRODUCT               1.22

POWER COST,U.S. $/METRIC T OF PRODUCT                0.20


RELEVANT VARIABLE COSTS,U.S. $/METRIC T OF PRODUCT   1.41

EXHAUST GAS FLOW AT TAOUT,M3/H                       1305

WEIGHT RATIO SOLIDS/GAS                              1.9



DRYING AIR TEMPERATURE,DEGREES C                    700
EXHAUST GAS TEMPERATURE,DEGREES C                    166
EVAPORATION,KG/H                                      60

EFFICIENCY,KJ/KG EVAPORATED                         14263

DIAMETER,M                                            0.20

MEDIUM COST,U.S. $/METRIC T OF PRODUCT               1.33

POWER COST,U.S. $/METRIC T OF PRODUCT                0.19


RELEVANT VARIABLE COSTS,U.S. $/METRIC T OF PRODUCT   1.53

EXHAUST GAS FLOW AT TAOUT,M3/H                       1285

WEIGHT RATIO SOLIDS/GAS                              2.0



DRYING AIR TEMPERATURE,DEGREES C                    800
EXHAUST GAS TEMPERATURE,DEGREES C                    185
EVAPORATION,KG/H                                      60

EFFICIENCY,KJ/KG EVAPORATED                         15521

DIAMETER,M                                            0.20

MEDIUM COST,U.S. $/METRIC T OF PRODUCT               1.45

POWER COST,U.S. $/METRIC T OF PRODUCT                0.19


RELEVANT VARIABLE COSTS,U.S. $/METRIC T OF PRODUCT   1.64

EXHAUST GAS FLOW AT TAOUT,M3/H                       1279

WEIGHT RATIO SOLIDS/GAS                              2.1
```

References

C-E Raymond Division, Combustion Engineering, Inc. (1985, 1986). Private communication.

Froessling, N. (1938). On the evaporation of falling droplets, *Gerlands Beitr. Geophysik* (Leipzig), 52, 170–216 (in German).

Kröll, K. (1978). *Drying Technology* (Part 2), Springer-Verlag, Berlin (in German).

Meedom, H. (1977). A compact system for the combustion of filter cakes, *Zement-Kalk-Gips* [Cement-Limestone-Gypsum], 30, 369 (in German).

Noden, D. (1972). Trend towards use of dispersion dryers, *Chem. Process Eng.*, 53, 48.

Noden, D. (1974). Efficient energy utilization in drying, *Processing*, December, 25.

Nonhebel, G. and A. A. H. Moss (1971). *Drying of Solids in the Chemical Industry*, Butterworths, London.

Perry, R. H., D. W. Green, and J. O. Maloney (1984). *Perry's Chemical Engineers' Handbook*, McGraw-Hill, New York.

Williams-Gardner, A. (1971). *Industrial Drying*, Leonard Hill, London, England.

8

SPRAY DRYING

8.1 Introduction

In Section 8.2, a general description of spray drying is given, and subsequent sections deal with droplet size prediction with rotary atomization, bulk density control of spray-dried products, and the possibilities of particle size enlargement. Particular attention is paid in Section 8.6 to crystallization heat because this feature can play a role in spray drying. Section 8.7 covers the recovery of the powder, and Section 8.8 the product transport by pneumatic conveying. Design methods are dealt with in Section 8.9. Section 8.10 explains ballpark data on pricing as is provided in Table 8.1, followed in Section 8.11 by a worked example. The chapter concludes with a discussion of the computer program given in Appendix 8.1 for the design of a spray dryer.

8.2 General description

Spray drying equipment accepts a feed in the fluid state (solution, suspension, or paste) and converts it into a dried particulate form by spraying the fluid into a warm or hot drying medium (usually air).

There are four principal stages in the spray-drying process (see Fig. 8.1): (a) feed atomization, (b) free moisture evaporation, (c) bound moisture evaporation, and (d) product recovery (air cleaning).

Spray drying produces relatively uniform spheres having the same proportion of nonvolatile components as the homogeneous liquid feed. Hence, it may be used to prepare complex mixtures of solids that cannot be produced

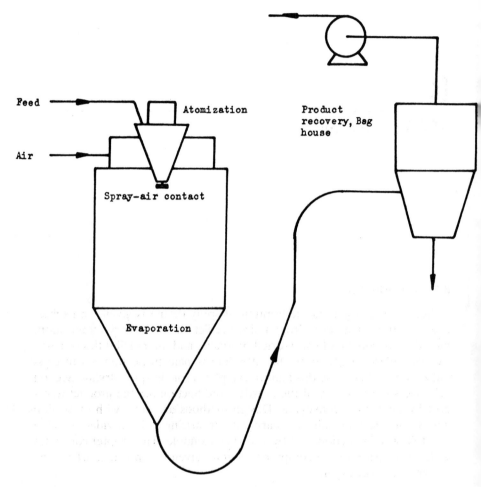

Figure 8.1   The process stages of spray drying.

by mechanical methods. For example, ceramic powders that are pressed into
spark plugs contain soluble organic binders and lubricants, and it is essential
that each dried particle has the same composition.

Spray drying may proceed cocurrently, countercurrently, or as a mixed-
flow process (see Fig. 8.2). Cocurrent drying exposes the droplets to the highest air temperature and hence, rapid evaporation results. Thus, this feature
may lead to products that have a low bulk density and consist of hollow struc-
tures. Countercurrent drying exposes particles that are almost dry to the highest temperatures and, hence, very dry products can be produced.

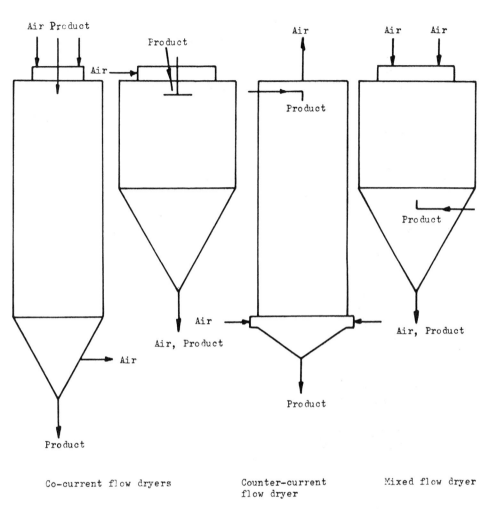

Figure 8.2 Product- and airflow modes.

It is impossible to both dry and cool in the same equipment; thus, cooling is often effected in a subsequent pneumatic system.

As a technique, spray drying dates back to the 1920s when it was first introduced into the milk and the detergent industries. With regard to scaling up, there is no limit, and present maximum feed rates are well over 100 tons per hour. However, this capacity does lead to very high capital costs when compared with other drying techniques.

There are three different devices that can be used to atomize the feed: (a) the single-fluid nozzle, (b) the pneumatic nozzle, and (c) the wheel (rotary atomizer).

Generally, the rotary atomizer (see Fig. 8.3) is preferred for large feed streams (i.e., exceeding 5 t/hr). It produces relatively small particles (30–120 μm) and the clogging tendencies are negligible on account of the large flow ports. The pneumatic (or two fluid) nozzle is used for small drying operations and sometimes with viscous feeds.

The single-fluid nozzle is used widely and produces larger particles than the wheel atomizer (e.g., 120–250 μm). However, the particle size is dependent on the feed pressure (between 50 and 300 atm), and, there are clogging tendencies.

A rotary atomizer can be used only in a cocurrent dryer whereas a nozzle can be used in cocurrent, countercurrent, and mixed-flow systems.

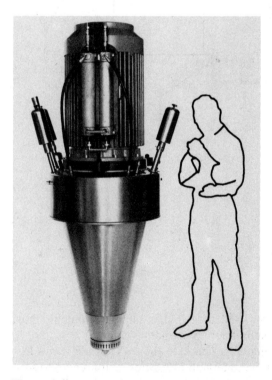

Figure 8.3 Modern rotary atomizer meeting the hygienic requirements of dairy, food, and pharmaceutical spray drying (Courtesy of Niro Atomizer, Copenhagen, Denmark).

In industry, the most common type of spray dryer is the open-cycle cocurrent system. A closed-cycle system will economize on energy, since part of the exhaust gas is recycled and the air-outlet temperature must be 1–3 K higher than in an open-cycle system to attain an identical final moisture content.

Usually, a spray dryer equipped with nozzle atomization has a cylindrical height three to four times larger than the chamber diameter, whereas the cylindrical height of dryers with rotary atomization is just about equal to the chamber diameter. This larger diameter prevents the spray from hitting the wall.

Spray dryers can be fired directly, using oil or gas; however, they are often used to process food (milk powder, coffee, etc.) and, consequently, direct firing is not used in these instances because of the nitrosamine issue and also because other nitrogen-containing compounds can be introduced, which could contaminate the food. Indirect heating by means of steam or electricity can be utilized.

The majority of spray dryers have air-inlet temperatures of less than 350°C, but there are spray dryers, processing inorganic materials (e.g., catalysts), which operate close to 800°C.

An interesting development recently is the combination of a spray dryer with a fluid-bed dryer. This equipment permits the lowering of the exhaust-gas temperature since the process stream can receive a posttreatment in a continuous fluid-bed dryer providing residence time.

Another recent development is the use of spray dry absorbers (SDA) whereby fluids containing toxic gaseous components are cleansed during a single pass through the SDA.

8.3 Droplet size control with rotary atomizers

Masters (1985) calculated the median droplet diameter in microns:

$$d_{50} = \frac{K \cdot M^a}{N^b \cdot d_i^c (n \cdot h)^d} \times 10^4 \, \mu m \tag{8.1}$$

And: $d_{95} = 2 \times d_{50} \ (d_{50} < 60 \, \mu m)$
$d_{95} = 2.5 \times d_{50} \ (d_{50} \sim 60 - 120 \, \mu m)$.

The median diameter defines a midpoint in the distribution, where half the total number of particles are smaller than the median and half are larger. Table 8.1, which can be used to select the power- and the K-values; shows that the average particle size is relatively independent of the feed rate and the wetted perimeter since the values of the exponents a and d are relatively low.

Table 8.1 Power- and K-values for Use in Equation (8.1)

Range of		Power value				
Wheel peripheral speed (m/sec)	Vane liquid loading (kg/hr · m)	a for feed rate	b for wheel speed	c for wheel diameter	d for wetted perimeter	K value
Normal 85–115	low 250	0.24	0.82	0.6	0.24	1.4
Normal-high 85–180	normal 250–1,500	0.2	0.8	0.6	0.2	1.6
very high 180–300	normal-high 1,000–3,000	0.12	0.77	0.6	0.12	1.25
normal-high 85–140	very high 3,000–60,000	0.12	0.8	0.6	0.12	1.2

The relationship is valid for vaned wheels. Masters (1985) gave two worked examples.

A satisfactory formula is not available for industrial nozzle operations (Masters, 1985). However, it has been proved empirically that a high pressure leads to a fine product and that a large orifice favors relatively large particles.

8.4 Bulk density control

The bulk density of a spray-dried product is often an important dependent process variable. The independent process variables that influence this property will be treated in succession:

Drying-gas inlet temperature

In cocurrent drying, temperature control is vital since contact between the hot gas and the droplet will lead to one of the possibilities that are indicated in Figure 8.4.

Masters (1985) gave a pronounced example of such dependence in experiments with sodium silicate (waterglass), using a SiO_2/Na_2O ratio of 2:1.

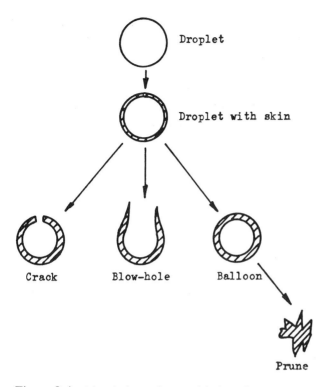

Figure 8.4   Morphology of spray-dried products.

Inlet temperature (°C)	Bulk density (kg/m^3)
500	70–100
250	250
180	500
150	650

Feed concentration

An increase in the feed concentration generally leads to an increase in the bulk density.

Solids content

Solid particles in the feed tend to lead to high bulk densities. Sometimes wet milling is effective in promoting this influence. Instances are known where spray drying of warm solutions yields a product consisting of hollow spheres having a bulk density of 320 kg/m^3, however, the processing of the same material at a lower temperature (the flow now containing crystals) exhibits a bulk density of 550 kg/m^3.

Aeration

A feedstock containing 50% air by volume yielded a product with a bulk density of 370 kg/m^3, with the particles containing vacuoles. A de-aerated feedstock led to 550 kg/m^3.

Feed temperature

This temperature variable is system dependent and can have a positive or negative influence.

Atomization

Wheels tend to lead to aeration but nozzles do not. Wheel constructions that reduce air-pumping effects (e.g., curved vanes) lead to products with higher bulk densities. The blanketing of wheels with live steam is an option worth considering.

Solid handling

A transport system that causes breakage will increase the bulk density.

Finally, there is a dependent process variable that strongly influences the bulk density; namely, the particle-size distribution. First, a fine material generally has a low bulk density because of the greater air volume between the particles. Measurements of the bulk density of both a milled and an unmilled crystalline organic powder show the following:

Average particle size (μm)	Bulk density (de-aerated) (kg/m^3)
10	330
20	625
150	850

Second, a material with a wide size distribution often has a high bulk density because the voids between the larger particles are filled by smaller particles.

8.5   Particle size enlargement

A spray dryer can be used for particle size enlargement when the particulate material is suspended in water with a binding agent. The binding agent leads to a type of suspension agglomeration (see Chapter 2). Subsequent spray drying produces agglomeration. Usually this process has an impact on the bulk density. Some commonly used binding agents are methyl cellulose, hydroxyethyl cellulose, and polyvinyl alcohol.

8.6 Heat of crystallization

A considerable amount of material is often crystallized during spray drying. Usually, heat is liberated during crystallization, although the reverse is also encountered especially with solutes exhibiting an inverted solubility characteristic. The crystallization of hydrates is often an endothermic process.

The heat of crystallization is taken as being equal in magnitude but opposite in sign to the heat of solution. Although this simplification is not entirely correct, the error involved is small. Reported heat-of-solution values refer to dissolution in an infinite amount of solvent; consequently, the heat of dilution is neglected. Heats of solution are usually reported at 18 or 25°C.

Example

The heat of solution of sodium chloride is -0.93 kcal/mol (the minus sign indicating an endothermic process). Therefore, the crystallization of a gmole of NaCl liberates 0.93 kcal. Since the molecular weight of sodium chloride is 58.46, the crystallization of 100 kg of NaCl liberates

$$\frac{100 \times 10^3 \times 4.19 \times 0.93}{58.46} = 6,666 \text{ kJ}$$

The specific heat of sodium chloride is 0.88 kJ/kg · K and, therefore, 6,666 kJ will raise the temperature of 100 kg of NaCl by:

$$\frac{6,666}{100 \times 0.88} = 76 \text{ K}$$

8.7 Product recovery

Since the spray dried product is invariably fine and the spent drying air from the dryer usually carries the total powder load, the selection of product recovery equipment is very important. One may distinguish between dry and wet product recovery. For dry product recovery, three principal choices exist: (a) dry cyclones (down to 5–10 micron), (b) bag filters (down to 0.1–0.01 micron), and (c) electrostatic precipitators (down to 1 micron).

A spray dryer is always equipped with a dry product-recovery system, cyclones and bag filters being the most popular choices. The average value of the gas face velocity is taken as 2 m/min for bag-filter design purposes.

The application of a wet product-recovery system depends upon specific requirements, wet scrubbers being used mainly for air cleaning. Although there is a great variety of equipment available, venturi scrubbers are very often used. Masters (1985) stated that a 95% collection efficiency for particles above 1 μm can be obtained. Pressure drops are usually of the order of 12–25 in wg. Normally, the slurry obtained is recycled.

8.8   Product transportation

Because of the small particle sizes being dealt with, pneumatic conveying is quite popular for spray-dried products. A typical value for the powder to air weight ratio is 4. It may be necessary to condition the air for the transport to prevent the powder from absorbing moisture from the transport air. Such conditioning can be relatively simple by cooling the air to, e.g., 5°C, whereby the water that is contained by the air condenses. The relatively dry air can then be used for transport. As a rule of thumb, line velocities between 20 and 25 m/sec are chosen.

Because the spray-dried material is relatively fine, it is feasible to combine the transport and the cooling of the powder. The heat transfer from the small particles is very rapid (see Chapter 7). However, in that case, the air mass flow should probably be boosted to obtain a powder-to-air mass ratio of approximately 1. Of course, this will only promote the reliability of the pneumatic transport but will add to the cost of the pneumatic transport equipment. If the powder-to-air mass ratio is approximately 1, the pressure loss equals approximately the pressure loss for clean air.

8.9   Design methods

The mass balance regarding the product is made first (see Chapter 4). The drying-gas exit temperature must be known to be able to calculate the heat transferred to the product flow, $Qtot_1$. The proposed relationship between TA_{IN} and TA_{OUT} is $TA_{OUT} = 88.39 \times \log_{10} TA_{IN} - 112.35$. Figure 8.5 graphically depicts the indicated points that were found in the literature (Kamenkovich et al., 1983; Kröll, 1978; Masters, 1985; Noden, 1972, 1974 and Perry et al., 1984) and observed for industrial spray dryers.

The data (65 sets of TA_{IN}/TA_{OUT} values) were analyzed statistically. The aforementioned relationship holds (linear regression), with the correlation coefficient being 0.727.

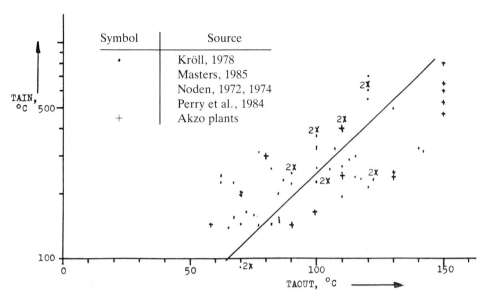

Figure 8.5 Relationship between air-inlet and air-outlet temperatures for spray dryers.

Apparently, there is a considerable spread in the data. A low TA_{OUT} leads to a good thermal efficiency; however, it is possible that the product still contains some moisture. A relatively high TA_{OUT} favors a low residual-moisture content. It is possible to give spray-dried material a posttreatment in a fluid-bed dryer. In that case TA_{OUT} can be relatively low.

To allow for the steady-state heat losses, $Qtot_1$ is multiplied by 1.25. Literature data on primary and secondary energy consumption of dryers are scarce. Three references are (Masters, 1985; Noden, 1972, 1974). Plant data were collected and these data together with the literature data led to the factor 1.25. The load factors to arrive at the performance in the long run are 1.1 (full), 1.2 (normal), and 1.3 (low), respectively. These figures are based on a small number of literature data and data of 5 Akzo spray dryers. It must be pointed out, however, that important large operations are usually quite efficient. On the other hand, small minor equipment functioning intermittently exhibits high specific consumption data.

It is assumed that a cocurrrent flow conical-based spray dryer (60° cone) with rotary atomization is selected. The volume of the dryer can be expressed as

$$V = 0.7854 \times D^2(H + 0.2886 \times D) \qquad m^3 \qquad (8.2)$$

H is the height of the cylindrical part and D the chamber diameter. A product residence time of 25 sec is a good design value. It is assumed that this residence time prevails if the chamber's volume exceeds the exit gas-flow in m³/sec by a factor of 25. The superficial air velocity is arrived at by dividing the exit gas flow in m³/sec by the chamber's cross-sectional area in m². Typical values for a spray dryer with rotary atomization are:

Chamber diameter, m	Superficial velocity, m/sec
4	0.15
6.5	0.35
9	0.55

The purpose is twofold: first, to have a cylindrical height being approximately equal to the diameter; second, to have the residence time of about 25 sec as a starting point.

One further aspect has to be observed. Dryers with rotary atomization may suffer from wall incrustations if the diameter is too small and the droplets are too large. As a rule of thumb, for a mean size between 80 and 100 μm, a minimum chamber diameter of 4 m is required (Masters, 1985).

Typical values for the superficial velocity in nozzle towers are much higher and vary between 1 and several m/sec. Here the cylindrical height normally exceeds the diameter by a factor of between 3 and 4. A cocurrent, open-cycle spray dryer can be equipped with a fan to transport the air to the tower, possibly through a direct air heater. A Δp of 2,500 N/m² is a fair approximation. An indirect heater (e.g., a steam heater) would require about 500 N/m².

An exhaust fan is often used to extract the gas/solids mixture from the spray chamber and can be located between the baghouse or the cyclone and a wet scrubber. A Δp of 1,500 N/m² is a good assumption. Another possibility is to have one exhaust fan to draw the gases through the complete system. The underpressure in the spray chamber usually exceeds the underpressure of a system equipped with two fans. A Δp of 4,000 N/m² appears to be a reasonable approximation of the actual state of affairs. An ingress air factor regarding the amount of fresh air is 1.1. Miscellaneous secondary energy consumption is 10 kW.

8.10   Investment aspects

Table 8.2 lists the variation of spray-dryer prices with regard to chamber size and evaporative capacity. The diameters are for chambers with a 60° conical base and a cylindrical height adequate to give an air residence time

Table 8.2 Spray-Dryer Prices

Chamber diameter m	Evaporative Capacity (kg of water per hr) at indicated TA_{IN}/TA_{OUT} in °C			Price $
	205/105	370/105	540/110	
3.8	170	430	650	190,000
5.0	430	1,090	1,640	275,000
5.6	705	1,755	2,630	310,000
6.4	1,115	2,780	4,160	395,000
7.8	1,790	4,420	6,610	525,000
8.5	2,890	7,095	10,600	650,000
9.1	3,615	8,880	13,255	825,000

Note: Based on cocurrent, open-cycle spray dryer with rotary atomizer, feed system, drying chamber, cyclone and scrubber, instrumentation, fan, motors, and ductwork (exclusive of supporting structure and penthouse).

of 25 sec. The diameter varies from 3.8 to 9.1 m. The drying-gas inlet temperatures are 205, 370, and 540°C, respectively. The dryer is equipped with a direct gas-fired heater. All parts that come into contact with the process streams are made from stainless steel. The table permits an estimate of a spray-dryer price by means of interpolation or extrapolation. The prices (as of 1985) were valid in the United States.

8.11   Worked example

Starting points

Product: Organic chemical
Airflow: Cocurrent
Atomization: Vaned wheel
Method of heating: Direct, natural-gas combustion
Initial moisture: 55% by wt
Final moisture: 0.5% by wt
Product rate: 500 kg/hr
Air-inlet temperature: 205°C
Feed temperature: 20°C
Ambient air temperature: 10°C

Literature data

Water specific heat: 4.19 kJ/kg · K
Water vapor specific heat: 1.886 kJ/kg · K
Latent heat of evaporation of water at 0°C: 2,504 kJ/kg · K

Water vapor specific mass relationship at 10^5 Pa: $220/(T + 273)$ kg/m^3
Air specific heat: 1.05 kJ/kg $\cdot$ K
Air specific mass relationship at 10^5 Pa: $355/(T + 273)$ kg/m^3
Organic chemical specific heat: 1.25 kJ/kg $\cdot$ K

The heat of crystallization is neglected.

Mass balance process flow, kg/hr

	In	Out
H_2O	608.1	2.5
Solids	497.5	497.5
	$\overline{}$ +	$\overline{}$ +
	1105.6	500

Evaporation 605.6 kg/hr

Heat-balance process flow

$TA_{OUT} = 88.39 \times \log_{10} 205 - 112.35 = 92.0°C$

$TP_{OUT} = 92.0 - 20 = 72°C$

$Q_1 = 605.6(2{,}504 + 1.886 \times 92.0 - 4.19 \times 20) = 1{,}570{,}752$
$Q_2 = 497.5 \times 1.25(72 - 20) \qquad\qquad = \quad 32{,}338$
$Q_3 = 2.5 \times 4.19(72 - 20) \qquad\qquad\quad = \qquad 545$

$Qtot_1 = \qquad\qquad\qquad\qquad\qquad\qquad\quad 1{,}603{,}635$ kJ/hr

$Qtot_2 = 1.25 \times \dfrac{(205 - 10)}{(205 - 92)} \times 1{,}603{,}635 = 3{,}459{,}168$ kJ/hr

Primary energy consumption $\dfrac{3{,}459{,}168}{605.6} = 5{,}712 \dfrac{kJ}{kg\ of\ H_2O}$

In the long run (full load) $1.1 \times 5{,}712 = 6{,}283 \dfrac{kJ}{kg\ of\ H_2O}$

$\dfrac{3{,}459{,}168}{32{,}000 \times 0.5} = 216$ nm^3 of natural gas/ton of product

In the long run: $1.1 \times 216 = 238$ nm^3/ton of product. Calorific value natural gas 32,000 kJ/nm^3.

Dryer size

Try a superficial gas velocity of 0.3 m/sec Quantity of air passed on to the dryer:

$\dfrac{3{,}459{,}168}{1.05(205 - 10)} = 16{,}895$ kg/hr

Quantity of air extracted from the dryer is $1.1 \times 16,895 = 18,585$ kg/hr (ingress air factor 1.1). $R_A = 0.973$ kg/m³ at 92°C.

$$\text{m}^3/\text{h at } 92°\text{C: } \frac{18,585}{0.973} = 19,101.$$

$$Evap = 605.6 \text{ kg/hr. } WA_{FLO} = \frac{605.6}{0.603} = 1,004 \text{ m}^3/\text{hr } (R_W = 0.603 \text{ kg/m}^3)$$

$Gasflo_2 = 19,101 + 1,004 = 20,105$ m³/hr at 92°C.

$$D = \sqrt{\frac{4 \times 20,105}{\pi \times 0.3 \times 3,600}} = 4.87 \text{ m}$$

Take 5.00 m. Take a residence time of 25 sec. Bottom: 60° cone.

$$V = \frac{25 \times 20,105}{3,600} = 139.6 \text{ m}^3$$

$H = 5.7$ m (which can be calculated by means of Equation (8.2))

Based on Table 8.2, an investment of $ 300,000 is at best a guess. With a Lang factor of 2.25, the total investment could then be: $2.25 \times 300,000 = \$ 675,000$.

Droplet size distribution

See Equation (8.1).
M = 1105.6 kg/hr
$d_1 = 0.25$ m
N = 10,000 rpm
n = 45
h = 30 mm
Wheel peripheral speed, m/sec: 131
Vane liquid loading, kg/hr · m: 819
Power values (Table 8.1) a: 0.2; b: 0.8; c: 0.6; d: 0.2
K: 1.6

$$d_{50} = \frac{1.6 \times 1105.6^{0.2} \times 10^4}{10,000^{0.8} \times 0.25^{0.6}(45 \times 0.03)^{0.2}} = 89 \ \mu\text{m}$$

$d_{95} = 2.5 \times 89 = 222 \ \mu$m

It is not necessary to reconsider the chamber diameter because of the mean droplet size. A minimum chamber diameter of 4 m is needed for 89 μm. Note: d_{50} is the median diameter (number basis).

Powder transport and cooling

It is preferred to cool the powder during transport.
A powder-to-air mass ratio of 1 is selected.
The line velocity is 25 m/sec.
Air specific heat: 1.0 kJ/kg · K.
Powder mass flow: 500 kg/hr.
Air mass flow: 500 kg/hr.

Calculation of the powder temperature

$500 \times 1.0(T - 10) = 500 \times 1.25(72 - T)$ kJ/hr
$T = 45°C$

Line diameter

$R_A = 1.25$ kg/m^3

$$\frac{500}{1.25} = 400 \text{ m}^3\text{/hr.} \qquad D = \sqrt{\frac{4 \times 400}{\pi \times 3600 \times 25}} = 0.075 \text{ m}$$

Bag filter

Face velocity 2 m/min.

$$\frac{450}{60 \times 2} = 3.8 \text{ m}^2$$

Take 5 m^2.

Electricity consumption

The dryer is equipped with two fans: an air supply fan and an exhaust fan.

Air-supply fan

Ambient air temperature: 10°C.

$R_A = 1.25$ kg/m^3 at 10°C.

$$Airflo_1 = \frac{16,895}{1.25} = 13,516 \text{ m}^3\text{/hr}$$

$$Pow_{G1} = \frac{13,516 \times 2,500}{3,600 \times 1,000 \times 0.5} = 18.8 \text{ kW}$$

Install a 25-kW motor.

Exhaust fan

Exhaust temperature: 92.0°C.

$R_A = 0.973$ kg/m^3 at 92.0°C.

$$Airflo_2 = \frac{16,895 \times 1.1}{0.973} = 19,100 \text{ m}^3/\text{hr}$$

$$R_W = 0.603 \text{ kg/m}^3 \text{ at } 92.0°C$$

$$WA_{FLO} = \frac{605.6}{0.603} = 1,004 \text{ m}^3/\text{hr}$$

$$Gas_2 = 19,100 + 1,004 = 20,104 \text{ m}^3/\text{hr}$$

$$Pow_{G2} = \frac{20,104 \times 1,500}{3,600 \times 1,000 \times 0.5} = 16.8 \text{ kW}$$

Install a 25-kW motor.

Solids-transport fan

$$\frac{400 \times 2,000}{3,600 \times 1,000 \times 0.5} = 0.4 \text{ kW}$$

Install a 1-kW motor.

Atomizer

$$8 \times 1.106 = 8.8 \text{ kW (8 kWhr/t of feed)}$$

Install a 12.5-kW motor.

Electricity consumption figures (kW)	
Air supply fan	18.8
Exhaust fan	16.8
Transport fan	0.4
Atomizer	8.8
Miscellaneous	10.0 +
	54.8

$$\frac{54.8}{0.5} = 109.6 \text{ kWhr/ton of product}$$

$$1.1 \times 109.6 = 121 \text{ kWhr/ton of product in the long run.}$$

Product recovery

A bag filter is selected, face velocity 2 m/min. No. of m² required:

$$\frac{20,104}{60 \times 2} = 167.5$$

Take 170 m².

Review

Dryer diameter, m: 5
Dryer cylindrical height, m: 5.7
Average droplet size, μm: 89
Gas consumption in the long run, nm^3/t of product: 238
Electricity consumption in the long run, kWhr/ton of product: 121
No. of m^2 filter area: 170

Drying costs

Fixed costs because of the investment (including maintenance).
Take an annual production of $7,000 \times 0.5 = 3,500$ t.
Total investment $675,000.

$$\frac{0.35 \times 675,000}{3,500} = \$67.50 / t$$

Gas costs
Basis: 0.1 $/nm^3
238 nm^3/t (in the long run)
$238 \times 0.1 = \$ 23.80/t$

Electricity costs
Basis: 0.06 $/kWhr
121 kWhr/ton (in the long run)
$121 \times 0.06 = \$ 7.26/t$

This picture does not take personnel costs into account. The same statement can be made for aspects like instrument air.

8.12 Computer program

Appendix 8.1 contains a computer program.
Input: lines 60–160.
Mass-balance process flow: 180–220.

A FOR-NEXT loop starts with line 250. The superficial velocity is gradually increased. A cylindrical height is calculated for each superficial velocity value. A residence time of 25 sec is a starting point. Thus, it is possible to select a velocity at which the diameter and the cylindrical height are approximately equal.

Line 270 calculates the drying-gas outlet temperature as a function of the selected drying-air inlet temperature.

Lines 280–310 concern the net heat (heat theoretically required by the process flow to have the product transformation performed). The product temperature's backlog is 20 K.

The heat input by the burner is given by line 320. The heat-loss factor (conduction, radiation, and convection losses) is 1.25. The chamber diameter and the cylindrical height are calculated by means of lines 330–470. The burner heat input is used to calculate the exit airflow (the exit water vapor flow being excluded) on line 350. An ingress air factor of 1.1 is used.

The 25-sec residence time emerges on line 430. V is the chamber volume. The primary energy consumption is calculated by lines 520 and 530. The secondary consumption (electricity) follows in lines 550–640. The calculations proceed as outlined in Section 8.10. The load factor is introduced in line 70.

The output appears by means of lines 710–890.

The program is followed by a processing of the example of Section 8.10. Equipment used: IBM Personal computer XT and printer Epson FX-80.

Notation

A_1	Feed moisture content	% by wt
A_2	Product moisture content	% by wt
Air_1	Airflow taken in by the dryer	m^3/hr
Air_2	Airflow taken in by the dryer	
Air_3	Airflow taken in by the dryer	
$Airflo_1$	Airflow taken in by the dryer	m^3/hr
$Airflo_2$	Airflow leaving the dryer	m^3/hr
a	Exponent in Equation (8.1)	
b	Exponent in Equation (8.1)	
Cal_{VAL}	Fuel or steam calorific value	mJ/t or mJ/nm^3
Cap	Product mass flow	kg/hr
CP	Solid specific heat	$kJ/kg \cdot K$
c	Exponent in Equation (8.1)	
D	Spray-dryer chamber diameter	m
	Line diameter	m
D_1	Spray-dryer chamber diameter	m
D_2	Spray-dryer chamber diameter	
D_3	Spray-dryer chamber diameter	
$Diam$	Spray-dryer chamber diameter	m
d	Exponent in Equation (8.1)	
d_1	Atomizing wheel diameter	m
d_{50}	Median diameter (number basis)	μm
d_{95}	Particle size below which 95% of the number of particles can be found	μm

Effic	Dryer efficiency	kJ/kg
Evap	Dryer water-evaporation load	kg/hr
Gas_2	Gas flow leaving the dryer	m^3/hr
Gas_3	Gas flow leaving the dryer	
Gas_4	Gas flow leaving the dryer	
$Gasflo_2$	Gas flow leaving the dryer	m^3/hr
H	Spray-dryer chamber cylindrical height	m
H_1	Spray-dryer chamber cylindrical height	m
H_2	Spray-dryer chamber cylindrical height	
H_3	Spray-dryer chamber cylindrical height	
h	Integer in FOR-NEXT loop	
	Atomizer wheel vane height	m
K	Proportionality constant in Equation (8.1)	Variable
Lofa	Load factor	
M	Atomizer wheel feed	kg/hr
Medco	Medium cost	$/$nm^3$ or $/t
Medcos	Medium cost per metric ton of product	$/t
Medflo	Medium flow	nm^3/hr or t/hr
N	Atomizer wheel rotational speed	min^{-1}
n	Number of vanes in the atomizer wheel	
Pow	Power consumption	kW
Pow_{ATOM}	Rotary atomizer power consumption	kW
Pow_{CO}	Power cost	$/kWhr
Pow_{COS}	Power cost per metric ton of product	$/t
Pow_{G1}	Fan power required for Air_1	kW
Pow_{G2}	Fan power required for Gas_2	kW
Q_1	Heat flow to the evaporated water	kJ/hr
Q_2	Heat flow for the dry solids	kJ/hr
Q_3	Heat flow for the residual water	kJ/hr
$Qtot_1$	Net heat	kJ/hr
$Qtot_2$	Steady-state dryer-heat requirement	kJ/hr
R_A	Air specific mass	kg/m^3
R_W	Water vapor specific mass	kg/m^3
Sol	Dry solids flow	kg/hr
T	Temperature	°C
TA_{IN}	Drying air temperature	°C
TA_{OUT}	Spent drying-gas temperature	°C
Tf	Feed temperature	°C
TP_{OUT}	Dried product temperature	°C
U	Superficial gas velocity	m/sec
U_{MAX}	Maximum superficial gas velocity	m/sec

V	Drying-chamber volume	m^3
Var_{COS}	Relevant variable cost per metric ton of product	\$/t
WA_{FLO}	Evaporated water flow	m^3/hr
W_{IN}	Water flow to the dryer	kg/hr
W_{OUT}	Water flow leaving the dryer	kg/hr
Δp	Pressure loss	N/m^2

Appendix   8.1

```
LIST
10   WIDTH "LPT1:",132
20   LPRINT CHR$(15);
30   LPRINT "SPRAY DRYER SCREENING PROGRAM"
40   LPRINT
50   LPRINT
60   INPUT "KG/H OUT";CAP
70   INPUT "LOAD FACTOR;FULL:1.1,NORMAL:1.2,LOW:1.3";LOFA
80   INPUT "FEED TEMPERATURE,DEGREES C";TF
90   INPUT "DRYING GAS TEMPERATURE,DEGREES C";TAIN
100  INPUT "MOISTURE CONTENT FEED,% BY WT";A1
110  INPUT "MOISTURE CONTENT PRODUCT,% BY WT";A2
120  INPUT "MAXIMUM SUPERFICIAL VELOCITY,h*.05,h IS AN INTEGER,M/S";UMAX
130  INPUT "SOLID SPECIFIC HEAT,KJ/(KG*K)";CP
140  INPUT "MJ/METRIC T:STEAM,OIL;MJ/NM3:GAS";CALVAL
150  INPUT "U.S.$/METRIC T:STEAM,OIL;U.S.$/NM3:GAS";MEDCO
160  INPUT "POWER COST,U.S.$/KWH";POWCO
170  LPRINT
180  REM MASS BALANCE IN KG/H
190  WOUT=.01*CAP*A2
200  SOL=CAP-WOUT
210  WIN=SOL/(100-A1)*A1
220  EVAP=WIN-WOUT
230  LPRINT
240  REM STUDY OF THE SUPERFICIAL VELOCITY EFFECT
250  FOR U=.1 TO UMAX STEP .05
260    REM HEAT BALANCE IN KJ/H
270    TAOUT=88.39*LOG(TAIN)/2.303-112.35
280    Q1=EVAP*(2504+1.886*TAOUT-4.19*TF)
290    Q2=SOL*CP*(TAOUT-20-TF)
300    Q3=WOUT*4.19*(TAOUT-20-TF)
310    QTOT1=Q1+Q2+Q3
320    QTOT2=QTOT1*(TAIN-10)/(TAIN-TAOUT)*1.25
330    REM DRYER SIZING IN M
340    RA=355/(TAOUT+273)
350    AIRFLO2=QTOT2/((TAIN-10)*1.05*RA)*1.1
360    RW=220/(TAOUT+273)
370    WAFLO=EVAP/RW
380    GAS2=AIRFLO2+WAFLO
390    D1=SQR(GAS2/(U*ATN(1)*3600))
400    D2=D1/.1
410    D3=INT(D2)+1
420    DIAM=.1*D3
430    V=GAS2/3600*25
440    H1=V/(.7854*DIAM^2)-.2886*DIAM
```

```
450    H2=H1/.1
460    H3=INT(H2)+1
470    H=.1*H3
480    GAS3=GAS2/1000
490    GAS4=INT(GAS3)+1
500    GASFLO2=GAS4*1000
510    REM HEATING MEDIUM COST IN U.S.$ PER METRIC T
520    MEDFLO=QTOT2/(CALVAL*1000)
530    MEDCOS=MEDFLO*MEDCO/CAP*1000*LOFA
540    EFFIC=MEDFLO*CALVAL*1000/EVAP*LOFA
550    REM POWER COST IN U.S.$ PER METRIC T
560    AIR1=QTOT2/(((TAIN-10)*1.05*1.24)
570    AIR2=AIR1/1000
580    AIR3=INT(AIR2)+1
590    AIRFLO1=AIR3*1000
600    POWG1=AIR1*2500/(3600*.5*1000)
610    POWG2=GAS2*1500/(3600*.5*1000)
620    POWATOM=(CAP*EVAP)/1000*8
630    POW=POWG1+POWG2+POWATOM+10
640    POWCOS=POW*POWCO/CAP*1000*LOFA
650    REM ADDING COSTS IN U.S.$ PER METRIC T
660    VARCOS=MEDCOS+POWCOS
670    REM OUTPUT
680    LPRINT
690    LPRINT
700    LPRINT
710    LPRINT USING "DRYING AIR TEMPERATURE,DEGREES C              ####";TAIN
720    LPRINT USING "EXHAUST GAS TEMPERATURE,DEGREES C             ###";TAOUT
730    LPRINT USING "EVAPORATION,KG/H                             ####";EVAP
740    LPRINT
750    LPRINT USING "EFFICIENCY,KJ/KG EVAPORATED                 #####";EFFIC
760    LPRINT
770    LPRINT USING "SUPERFICIAL VELOCITY,M/S                      .##";U
780    LPRINT USING "DIAMETER,M                                  ##.#";DIAM
790    LPRINT USING "CYLINDRICAL HEIGHT,M                        ##.#";H
800    LPRINT
810    LPRINT USING "MEDIUM COST,U.S.$/METRIC T OF PRODUCT       ###.##";MEDCOS
820    LPRINT
830    LPRINT USING "POWER COST,U.S.$/METRIC T OF PRODUCT        ###.##";POWCOS
840    LPRINT
850    LPRINT USING "RELEVANT VARIABLE COSTS,U.S.$/METRIC T OF PRODUCT   ###.##";VARCOS
860    LPRINT
870    LPRINT USING "INLET AIR FLOW AT 10 DEGREES C,M3/H         #####";AIRFLO1
880    LPRINT
890    LPRINT USING "EXHAUST GAS FLOW AT TAOUT,M3/H              #####";GASFLO2
900    LPRINT
910    NEXT U
920 LPRINT
930 WIDTH "LPT1:",80
940 LPRINT CHR$(18)
950 END
0


RUN
SPRAY DRYER SCREENING PROGRAM
```

```
KG/H OUT? 400
LOAD FACTOR;FULL:1.1,NORMAL:1.2,LOW:1.3? 1.2
FEED TEMPERATURE,DEGREES C? 50
DRYING GAS TEMPERATURE,DEGREES C? 220
MOISTURE CONTENT FEED,% BY WT? 55
MOISTURE CONTENT PRODUCT,% BY WT? 1
MAXIMUM SUPERFICIAL VELOCITY,h*.05,h IS AN INTEGER,M/S? .3
SOLID SPECIFIC HEAT,KJ/(KG*K)? 1.2
MJ/METRIC T:STEAM,OIL;MJ/NM3:GAS? 32
U.S.$/METRIC T:STEAM,OIL;U.S.$/NM3:GAS? .1
POWER COST,U.S.$/KWH? .06
```

```
DRYING AIR TEMPERATURE,DEGREES C                       220
EXHAUST GAS TEMPERATURE,DEGREES C                       95
EVAPORATION,KG/H                                       480

EFFICIENCY,KJ/KG EVAPORATED                           6279

SUPERFICIAL VELOCITY,M/S                                .10
DIAMETER,M                                             7.0
CYLINDRICAL HEIGHT,M                                   0.5

MEDIUM COST,U.S.$/METRIC T OF PRODUCT                 23.54

POWER COST,U.S.$/METRIC T OF PRODUCT                   7.43

RELEVANT VARIABLE COSTS,U.S.$/METRIC T OF PRODUCT     30.97

INLET AIR FLOW AT 10 DEGREES C,M3/H                  10000

EXHAUST GAS FLOW AT TAOUT,M3/H                       14000
```

```
DRYING AIR TEMPERATURE,DEGREES C                       220
EXHAUST GAS TEMPERATURE,DEGREES C                       95
EVAPORATION,KG/H                                       480

EFFICIENCY,KJ/KG EVAPORATED                           6279

SUPERFICIAL VELOCITY,M/S                                .15
DIAMETER,M                                             5.7
CYLINDRICAL HEIGHT,M                                   2.2

MEDIUM COST,U.S.$/METRIC T OF PRODUCT                 23.54

POWER COST,U.S.$/METRIC T OF PRODUCT                   7.43

RELEVANT VARIABLE COSTS,U.S.$/METRIC T OF PRODUCT     30.97

INLET AIR FLOW AT 10 DEGREES C,M3/H                  10000

EXHAUST GAS FLOW AT TAOUT,M3/H                       14000
```

```
DRYING AIR TEMPERATURE,DEGREES C                        220
EXHAUST GAS TEMPERATURE,DEGREES C                        95
EVAPORATION,KG/H                                        480

EFFICIENCY,KJ/KG EVAPORATED                            6279

SUPERFICIAL VELOCITY,M/S                                .20
DIAMETER,M                                              5.0
CYLINDRICAL HEIGHT,M                                    3.5

MEDIUM COST,U.S.$/METRIC T OF PRODUCT                 23.54

POWER COST,U.S.$/METRIC T OF PRODUCT                   7.43

RELEVANT VARIABLE COSTS,U.S.$/METRIC T OF PRODUCT      30.97

INLET AIR FLOW AT 10 DEGREES C,M3/H                   10000

EXHAUST GAS FLOW AT TAOUT,M3/H                        14000




DRYING AIR TEMPERATURE,DEGREES C                        220
EXHAUST GAS TEMPERATURE,DEGREES C                        95
EVAPORATION,KG/H                                        480

EFFICIENCY,KJ/KG EVAPORATED                            6279

SUPERFICIAL VELOCITY,M/S                                .25
DIAMETER,M                                              4.5
CYLINDRICAL HEIGHT,M                                    4.8

MEDIUM COST,U.S.$/METRIC T OF PRODUCT                 23.54

POWER COST,U.S.$/METRIC T OF PRODUCT                   7.43

RELEVANT VARIABLE COSTS,U.S.$/METRIC T OF PRODUCT      30.97

INLET AIR FLOW AT 10 DEGREES C,M3/H                   10000

EXHAUST GAS FLOW AT TAOUT,M3/H                        14000




DRYING AIR TEMPERATURE,DEGREES C                        220
EXHAUST GAS TEMPERATURE,DEGREES C                        95
EVAPORATION,KG/H                                        480

EFFICIENCY,KJ/KG EVAPORATED                            6279

SUPERFICIAL VELOCITY,M/S                                .30
DIAMETER,M                                              4.1
CYLINDRICAL HEIGHT,M                                    6.1

MEDIUM COST,U.S.$/METRIC T OF PRODUCT                 23.54
```

POWER COST,U.S.$/METRIC T OF PRODUCT	7.43
RELEVANT VARIABLE COSTS,U.S.$/METRIC T OF PRODUCT	30.97
INLET AIR FLOW AT 10 DEGREES C,M3/H	10000
EXHAUST GAS FLOW AT TAOUT,M3/H	14000

References

Dittman, F. W. and E. M. Cook (1977). Establishing the parameters for a spray dryer, *Chem. Eng.*, 84, 108.

Kamenkovich, V. V., T. A. Solov'eva, I. D. Goikhman, S. I. Pad'ko, and V. E. Babenko (1983). Features of the industrial process of spray drying an intermediate detergent composition, *Sov. Chem. Ind.* 3, 364.

Kröll, K. (1978). *Drying Technology* (Part: 2), Springer-Verlag, Berlin (in German).

Masters, K. (1985). *Spray Drying Handbook*, George Godwin, London.

Noden, D. (1972). Trend towards use of dispersion dryers, *Chem. Process Eng.*, 53, 48.

Noden, D. (1974). Efficient energy utilization in drying, *Processing*, December, 25.

Perry, R. H., D. W. Green, and J. O. Maloney (1984). *Perry's Chemical Engineers' Handbook*, McGraw-Hill, New York.

9

MISCELLANEOUS CONTINUOUS DRYERS

9.1 Introduction

Whereas the "big four" (i.e., fluid bed, direct-heat rotary, flash, and spray dryers) were discussed in the preceding chapters, there still remain some important convective-type dryers to be dealt with in this chapter.

The conveyor dryer is indispensable for the gentle drying of relatively coarse particulate material, the somewhat large particle size normally being obtained by a forming machine. In the ceramic industry, the Hazemag Rapid dryer is well established for the processing of plastic masses. The Convex dryer resembles the flash dryer, however, it occupies less space and can provide residence time. A series of contact dryers is covered, their fundamental bonus being their modest energy consumption. In this respect, they are more efficient than convective dryers but they all have a limited capacity. Scaling up is not effective because the number of unit heat-transfer areas per unit of volume decreases. For example, equipment having triple the volume has only double the area. Plate dryers, agitated contact dryers (both mildly and vigorously agitated), vertical thin-film dryers, steam-tube dryers, and drum dryers are all considered in this chapter.

9.2 Conveyor dryers

General description

A conveyor dryer is designed so that material is dried as it is continuously being transported horizontally on a perforated screen through which warm air is blown. The most widely used type is the single-conveyor dryer (see

Fig. 9.1). Staging is employed when a large amount of shrinkage occurs (e.g., with the drying of onions) and repacking is necessary to promote material to air contact. Conveyors in series can have different speeds (see Fig. 9.2).

Multiple-conveyor dryers are used when, for quality reasons, drying must proceed with low air velocities and temperatures (e.g., for pastas; see Fig. 9.3).

While being dried, the material is stationary and consequently the technique is suitable for friable products. The bed of wet material must be permeable which means that the feed must be particulate. Since the required particle size is rather large (from 1 mm to several cm) often feed preforming is necessary (by, e.g., an extruder). It is very important to distribute the feed carefully (e.g., by means of an oscillating spreader) as there is no immediate opportunity to rearrange the bed of solids (compare, e.g., a fluid-bed dryer). Sticking of the particles to each other or to the apron cannot be tolerated.

Most conveyor dryers have zonal internal air recirculation. With zone or cell or a combination of cells, there is a fresh air make-up and a purge.

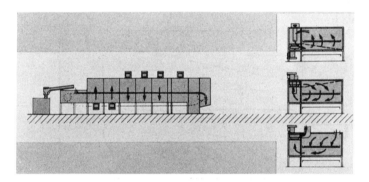

Figure 9.1 A single-conveyor dryer. (Courtesy of Proctor & Schwartz, Inc., Horsham, Pennsylvania.)

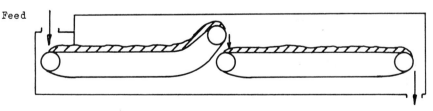

Figure 9.2 A two-stage conveyor dryer (seen schematically).

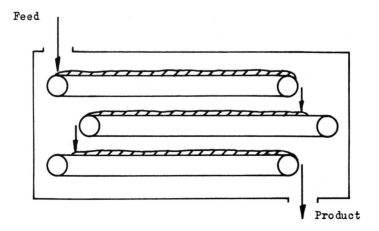

Figure 9.3   A multiple conveyor dryer (seen schematically).

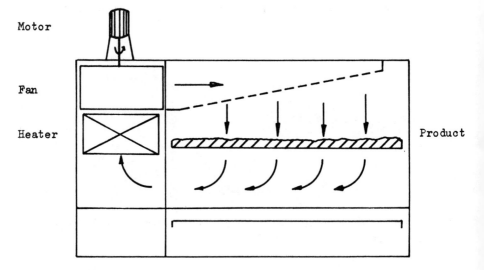

Figure 9.4   A cross-sectional view of a conveyor dryer cell.

Internal air recirculations pass the bed of material to be dried, and go through a fan and a heat source (usually a steam heater). Typically, the exhaust is controlled by the temperature in the loop (see Fig. 9.4).

A typical single-stage conveyor dryer consists of a number of almost identical cells in series between two elements having a different construction. An exhaust is active for a number of cells, and the same scheme applies for a make-up. It is easy to picture an airflow regime in which all warm air passes through once as in a fluid-bed dryer, which would make sense for a conveyor dryer, too, as long as the material is in the constant drying-rate period. However, materials that are being processed in a conveyor dryer soon enter the period of drying with a decreasing rate because of the relatively large shape of the particles and the porous nature of the material.

Operating on a "once-through" basis would mean that exhaust air with a low degree of saturation would be obtained. Usually, exhaust air has a relative humidity between 10 and 40%. The net effect is a saving of energy.

The last cell or last set of cells is often used for cooling the material down to 40°C at least. Atmospheric air is used; which here, of course, passes through once.

A practice that is sometimes adopted is to not only control the temperature in the recirculation loop but also the relative humidity by steam injection—the reason being to obtain a good product. An example is the drying of polyester chips that, if dried otherwise, are less suitable for a subsequent process step; in this case, dyeing.

Layer depth

On the average, depth varies between 2 and 15 cm, with 4 and 6 cm being quite common values. The pressure loss of the drying air that flows through the bed is related to the air velocity and the layer depth. This fact explains one reason to work with relatively shallow layers because pressure drops exceeding 25–50 mm wg would lead to leaks or to by-passing (e.g., through the slit between the moving belt and the casing).

Air velocity

Practical air velocities vary between 0.5 and 2 m/sec with 1.25 m/sec being a common value. Often an upward flow is chosen at the beginning of the drying period and a downward flow later on. This regime promotes even drying and avoids dusting.

Temperature

Most conveyor dryers have drying-air temperatures between 100 and 200°C with even higher temperatures a possibility.

Size

There is no practical limit with regard to the length of a conveyor dryer, and the width can vary between 0.5 and 3 m.

Conveyor velocity

Usually, the velocity is continuously adjustable between 0.1 and 2 cm/sec.

Pricing

Proctor & Schwartz (of Horsham, Pennsylvania) kindly provided two price indications regarding the cost of dryers (1987 quotation). First, approximately $2,500 per running foot for the length, including end extensions. This price is applicable for a narrow (under 4-foot wide conveyor) pilot machine made from aluminized and painted plain steel. Second, the price can be upwards of $5,000 per foot for a wide (over 11-feet wide conveyor) dryer made from 316 stainless steel with special design features. Figure 9.5 shows an installed and operating dryer.

Design method

The design of a conveyor dryer is based on the registration of drying curves. Representative wet material is deposited on a test tray (e.g., 0.3 × 0.3 m^2) with air blown through. Independent variables (no recycle) include: (a) air conditioning (temperature and RH), (b) air velocity, and (c) layer depth.

Dependent variables (as fuctions of time) include: (a) exhaust air temperature, (b) pressure loss across the material, and (c) sample weight. These curves allow the selection of the right air conditions, the air velocity,

Figure 9.5 An operating crouton conveyor dryer. (Courtesy of Proctor & Schwartz, Inc., Horsham, Pennsylvania.)

and the layer depth. These variables define the capacity per m² and per hr. A scaling-up factor of $1/0.7 = 1.42$ is normally adopted. The reasons are that the material is, in practice, distributed poorer and that the actual large-scale situation is comparable but not identical to the laboratory situation because of the zonal recirculation. In the laboratory, tests are performed on a once-through basis.

Attention is also paid to product shrinkage, fines generation, fume or smoke generation, tendency to stick to the test tray, and cooling of the product.

Sample calculation

Starting points

Capacity: 2,500 kg/hr out of the dryer
Initial moisture content: 30% by wt (wet basis)
Final moisture content: less than 0.5% by wt
Physical form: 5 mm diameter extruded pellets, length varies between 5 and 10 mm
Cooling: below 40°C required
Maximum product temperature: 90°C
Bulk density feed: 550 kg/m³
Solid specific heat: 1.2 kJ/kg · K

The product is neither toxic nor corrosive and the liquid to be evaporated is water. There is no generation of fines that might lead to a dust explosion.

Calculation

Mass balance, kg/hr

	In	Out
H_2O	1,066.1	12.5
dry matter	2,487.5 +	2,487.5 +
	3,553.6	2,500

Evaporation $1,066.1 - 12.5 = 1,053.6$

Drying curve   See Figure 9.6. Test tray 0.3×0.3 m².

Independent process variables that were selected are: (a) air temperature 90°C (no RH adjustment), (b) air velocity 1 m/sec, and (c) layer depth of 4 cm.

Dependent variables that were selected include: (a) 15 minutes to come down to 0.3% by wt, (b) mandatory raising of the product's temperature to 80°C, and (c) pressure drop of the bed is 10 mm wg.

Nominal conveyor width: 3.00 m
Drying length: $57.4/3 = 19.1$ m

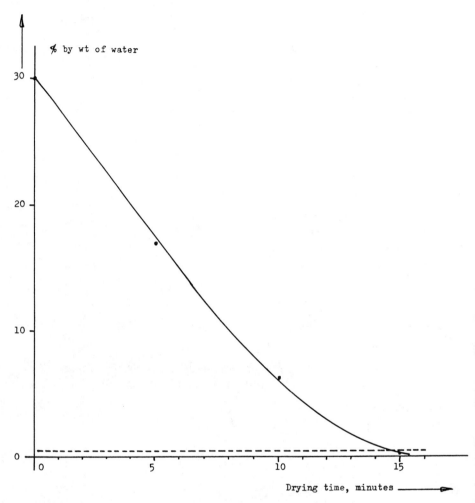

Figure 9.6   Drying curve of the sample calculation.

Test mass balance, kg

	In	Out
H_2O	0.59	0.00
Dry matter	$\underline{1.39}$ +	$\underline{1.39}$ +
	1.98	1.39

Capacity $1.39 \times 60/15 \times 1/0.09 = 61.8$ kg/m$^2 \cdot$ hr
Area required $1.42 \times 2{,}500/61.8 = 57.4$ m^2

Standard section length: 3.5 m
Number of sections required for drying: 19.1/3.5 = 5.45
Take 6 sections

Cooling curve   Test tray 0.09 m². Two minutes are required to get below 40°C with an air temperature of 20°C and a velocity of 1 m/sec.
Capacity: $1.39 \times 60/2 \times 1/0.09 = 463$ kg/m² · hr
Area required: 2,500/463 = 5.4 m²
One cooling section is adequate.

Investment   Take 316 stainless steel and a Lang factor of 2.25:

$$\frac{2.25 \times 7 \times 3.5 \times 5,000}{0.3048} \approx \$900,000 \text{ [U.S.]}$$

Heat requirement   The exhaust temperature will be 70°C. The product exit temperature is 80°C (inlet 20°C).

$$
\begin{aligned}
Q_1 &= 1,053.6(2,504 + 1.886 \times 70 - 4.19 \times 20) &&= 2,689,019 \\
Q_2 &= 2,487.5 \times 1.2(80 - 20) &&= 179,100 \\
Q_3 &= 12.5 \times 4.19(80 - 20) &&= \underline{3,143} + \\
& && \quad 2,871,262 \qquad \text{kJ/hr}
\end{aligned}
$$

Air make-up 15,000 m³/hr at 10°C and 50% RH (0.004 kg of H_2O/kg of dry air).

kg/hr

	In	Evaporation	Out
H_2O	75	1,053.6	1,128.6
Dry air	18,741 +	0 +	18,741 +
	18,816	1,053.6	19,869.6

Moisture content exhaust is 0.060 kg H_2O/kg dry air. RH is approximately 28%.

Air make-up is heated from 10°C to 70°C.

Heat requirement

$$
\begin{aligned}
75 \times 1.886(70 - 10) &= 8,487 \\
18,741 \times 1.00\ (70 - 10) &= \underline{1,124,460} + \\
& \quad\ \ 1,132,947 \qquad \text{kJ/hr}
\end{aligned}
$$

Note: this heat requirement increases considerably when the ambient-air temperature is −20°C.

Add: 2,871,262
      $\underline{1,132,947}$ +
      4,004,209   kJ/hr

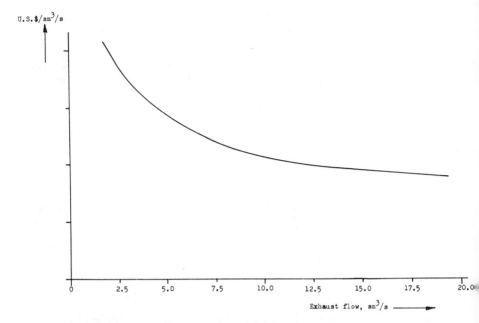

Figure 9.7   Average installed cost of heat-recovery systems (qualitatively). (Courtesy of Proctor & Schwartz, Inc., Horsham, Pennsylvania.)

Multiply by 1.2 to allow for conduction, convection, and radiation losses: 4,805,051 kJ/hr

Latent heat of evaporation of steam at 4 atm: 2,136 kJ/kg. 4,805,051/ 2,136 = 2,250 kg of steam/hr, say, 2.3 t/hr.

Heat recovery aspects   The dew point of the exhaust is approximately 43°C; hence, cooling from 70 to 50°C to warm air make-up indirectly is possible.

Figure 9.7 gives qualitatively the average installed cost of heat-recovery systems as a function of the exhaust flow in standard m^3 (15.6°C, atmospheric pressure) per sec.

	kg/hr	ρ, kg/m^3 at 15.6°C	sm^3/sec
H_2O	1,128.6	0.762	0.4
Dry air	18,741 +	1.230	4.2 +
	19,869.6		4.6

Annual savings for 7,000 hr per annum
7,000 × [1,128.6 × 1.886(70 − 50) + 18,741 × 1.00(70 − 50)] = 2,921,735,500 kJ per annum.
Metric tons of steam saved: 2,921,735,500 / (1,000 × 2,136) = 1,367.8.
At $12/ton: 12 × 1,367.8 = $16,414.

The investment required must be balanced against these savings. As the annual savings are relatively small, the prospects for a project along these lines are not particularly good.

Fixed costs and variable costs Dryer operates 7,000 hr per annum. Annual production: $7,000 \times 2.5 = 17,500$ metric ton

$$\text{Fixed costs per ton } \frac{0.35 \times 900,000}{17,500} = \$18$$

Steam costs per ton
Take an average steam consumption in the long run of $1.25 \times 2.3 = 2.9$ t/hr ($12/metric ton of steam)

$$\frac{2.9 \times 12}{2.5} = \$13.92 \text{ per metric ton of product}$$

9.3 Hazemag Rapid dryer

An example of an agitated continuous dryer, the Hazemag Rapid operates with a cocurrent flow of the feedstock and the drying gas, as depicted schematically in Figure 9.8. In Figure 9.9, an installed Rapid dryer is shown that processes industrial sludge (approximate dryer length is 6 m). The stationary dryer shell is provided with double pendulum self-sealing feed-

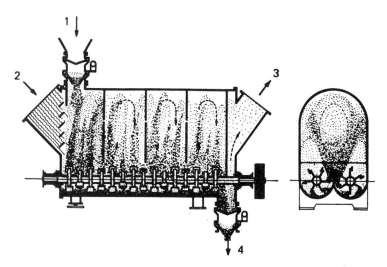

Figure 9.8 Scheme of a Rapid dryer. (1) Feed, (2) Air inlet, (3) Air outlet, and (4) Product. (Courtesy of Salzgitter Maschinenbau GmbH, Münster, Federal Republic of Germany.)

Figure 9.9   A Rapid dryer that is used for industrial sludge. (Courtesy of Salz-gitter Maschinenbau GmbH, Münster, Federal Republic of Germany.).

and dry-product connections which may be coated with teflon to prevent incrustations. One or a pair of rotating shafts carrying paddles disintegrate the feedstock and propel it through the dryer. A typical agitation system can be seen in Figure 9.10. The paddles are made of wear-resisting material and are easily replaceable. Typical peripheral speeds are in the range 6–12 m/sec. The one-shaft type may be selected for relatively small evaporative loads (maximum, e.g., 2.5 metric t/hr) if the drying duty is not too difficult (i.e., feed disintegration is fairly easy). The two-shaft type is taken for

Figure 9.10 A view into the interior of a Rapid dryer. (Courtesy of Salzgitter Maschinenbau GmbH, Münster, Federal Republic of Germany.)

larger jobs and for smaller ones if the disintegration has to be concentrated upon. Furthermore, the two-shaft type is usually required for low-temperature drying (using exhaust gases). The rotational directions are such that at the center a fountain is being created. The shaft bearings near the feeding point can be water-cooled. The agitation power input is in the range of 0.75–1.5 kWhr per metric ton of product.

The material of construction of the dryer is steel, with the paddles and the lower shell part being made of wear-resistant steel. However, stainless steel may also be used. Usually, the dryer is not insulated, for the reason that the relatively high inlet temperatures (600–800°C) would, in an insulated dryer, lead to too high temperatures for vital mechanical parts, such as bearings. The dryer is being used for mineral sludges, coal sludge, clay, and various filter cakes. The processing of plastic masses for the ceramics industry is particularly successful with this dryer.

The spent drying gas characteristically exits at temperatures of 100–120°C. Residence time of solids is between 2 and 6 sec. The evaporative loads are in the range of 50–25,000 kg/hr.

Typical gas velocities in the Rapid dryer can approach and exceed 12 m/sec. Such relatively high gas velocities imply a considerable carry-over of fines into the cyclones. A distribution of the product from the dryer and from the cyclones is 85/15.

Semi-technical experiments can be carried out in the smallest Rapid dryer, which has an approximate length of 2.5 m. This dryer is equipped with 2 shafts, and the results can be used for scaling up. The largest two-shaft dryer is about 7.6 m long and 5.4 m wide.

9.4   Convex dryer

The Convex dryer, which is a continuously functioning convective dryer, is schematically depicted in Figure 9.11. The drying gas containing the material to be dried enters the drying chamber tangentially. The rotational

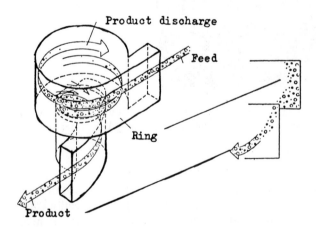

Figure 9.11   A Convex dryer (seen schematically). (Courtesy of Buss, Basel, Switzerland.)

Figure 9.12 A right-hand Convex dryer with concentric rings. (Courtesy of Buss, Basel, Switzerland.)

movement of the gas/solid system induces a centripetal force leading to a ring of particles close to the wall. The continuous make-up of particles causes a displacement of material to the center of the chamber.

Moreover, the large particles stay in the ring much longer than the small ones. The latter effect is desired because large particles require longer residence times than small ones, cf. Chapter 7, the chapter on flash drying.

The provision of internals in the drying chamber is a further development. Figure 9.12 illustrates the presence of two concentric rings promoting plugflow and long residence times.

The residence time of the particles is a function of: (a) the particle size, (b) the load factor (i.e., kg of solids per kg of drying gas), (c) the chamber geometry, and (d) the volumetric flow (giving rise to a certain circumferential velocity in a chamber with fixed dimensions).

Particle residence times of 10–20 sec can be expected for particle sizes up to 2 mm and a load factor of 0.3. The range 10–20 sec can be expected for different geometries. These times can be markedly increased by the provision of internals and by putting chambers in series (see Fig. 9.13).

The feeds must be relatively free-flowing and must not give rise to incrustations. In some instances, the installation of an agitator facilitated the processing of pasty products.

Important typical applications

1. Carboxy methyl cellulose and other cellulose ethers
2. S-PVC

Figure 9.13 Convex dryers in a series for gypsum calcining (brick-lining). (Courtesy of Buss, Basel, Switzerland.)

3. MBS (a PVC modifier)
4. Organic fungicides
5. Gypsum
6. Silica

Convex dryers were recently successfully applied to the calcination of catalysts, aluminium hydroxide and zeolites. Calcination implies the removal of chemically or physically bound water and usually proceeds at material temperatures between 500 and 1,200°C. A transition of crystalline modification No. 1 into crystalline modification No. 2 may proceed simultaneously.

Various lay-outs to calcine as economically as possible are proposed. The removal of free water commonly precedes the chemical reactions. Pilot plant tests are being carried out with a 0.56-m diameter dryer.

9.5   Plate dryer

The plate dryer is a continuous contact dryer and is depicted schematically in Figure 9.14. The feedstock is continuously fed into the top of the dryer. The product is conveyed in a spiral pattern across stationary plates by rakes

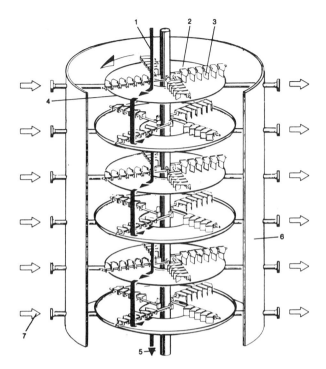

Figure 9.14 A plate dryer (seen schematically). (1) Wet product, (2) Plate, (3) Rabble mechanism, (4) Plate rim, (5) Dry product, (6) Housing, and (7) Heating or cooling medium. (Courtesy of Krauss Maffei, Munich, Federal Republic of Germany.)

on a vertical rotating shaft. The plates are heated or cooled by a medium. Even number plates have a larger diameter than odd number plates.

Both this arrangement and the positioning of the plows attached to the radial arms enable the processed material to take the path that is indicated. Note that this set-up guarantees plugflow. Each plate or group of plates can be heated or cooled individually. Evaporated water or solvent is removed by strip gas (air or nitrogen) in crossflow.

The feedstock enters via a rotating lock. It is also possible to dry continuously under vacuum. The relatively high price of the vacuum equipment may be more than compensated by a faster evaporation rate leading to a smaller dryer. The vacuum-tight locks for the product flow are an additional cost item.

With the use of a thermal oil, the plate temperature can be as high as 300°C. The shaft rotates typically at 2–6 rpm. The dryer can be applied in a number of cases. An important aspect is the low gas velocity leading to little if any dust carry-over. The processing of materials having a wide particle size distribution is possible. The size of the gas cleaning system is quite modest. A further aspect is the gentle treatment of the material dried. The dryer can provide residence time that is often required for the removal of firmly bound moisture (5–150 min).

Materials that are fed into the dryer (maximum capacity 10 m³/hr) must be free-flowing. Possible incrustation tendencies may be decreased by coating the plows with teflon.

Typical transverse gas velocities are between 0.1 and 0.2 m/sec. It is usual to work with shallow product layers (i.e., 10–15 mm). A characteristic figure is the product turn-over number which indicates the number of times a specific particle is "reshoveled" during its stay in the dryer. (This will be elucidated in the sample calculation.)

Experiments with a 0.1- or 0.15-m² plate dryer are usually adequate for scaling-up. In some cases testing on a 5-m² dryer is required. Essentially, the small dryer is one plate with an agitator.

Industrial dryers can have a diameter of 2 m or 2.7 m, and the number of plates can range from 5 to 35. Dryers with areas from 12.5 to 175 m² are thus available.

Sample calculation

A series of tests are carried out, with one test selected for scaling-up.

Process data
Plate temperature: 95°C
Air temperature: 95°C
Gas velocity: 0.2 m/sec
Agitator:
  Number of arms: 2
  Number of plows per arm: 2
  Rotational speed: 5 rpm

Mass balance product, g

	In	Out
H_2O	123	0
solids	492 +	492 +
	615	492

Test dryer size: 0.1 m²
Initial layer thickness 10 mm

Registrations
Drying time: 37 min
Final product temperature: 85°C
Product turn-over number: $2 \times 5 \times 37 = 370$

Scaling up
Industrially, an output of 350 kg/hr is desired.

Required area: $0.1 \times \dfrac{37}{60} \times \dfrac{350}{0.492} = 43.9 \text{ m}^2$

With an allowance for losses and cooling, a 50-m^2 dryer is selected (e.g., 20 plates of 2.5 m^2): TTB 20/20. A large-scale plate dryer is depicted in Figure 9.15.

Figure 9.15 A plate dryer TTB 20/11 (11 plates, nominal diameter 2 m) with dosing device. (Courtesy of Krauss Maffei, Munich, Federal Republic of Germany.)

A dryer resembling the plate dryer is the wiped-tray dryer. Here, too, the material cascades from tray to tray through the dryer. However, this is a convective dryer, with air being sucked into the dryer, then heated up and circulated through the dryer by fans. The off-gases disappear into the stack at the top. The plates are not heated.

9.6 Spiral dryer

The spiral dryer is a short-time contact dryer for particulate materials that do not stick and that can be conveyed pneumatically. Figure 9.16 depicts a spiral dryer schematically. (Both horizontal and vertical installation are possible.) The material to be dried and the conveying gas enter at the dryer's base. Air guide plates on the rotor that are interrupted and staggered create spiral-shaped channels through which the gas flows on entraining the moist material, which is thus conveyed as a thin film along the heated tube wall. Rapid heat transfer can take place during the contact time of several seconds. The moisture released from the powder is absorbed by the gas. The jacket and the rotor are heated.

Continuously reheating the carrier gas guarantees a great driving force for the moisture release all the way up to the exit. Because of the location of the air guide plates and the rotation (0–10 rpm) the formation of fixed flow-paths and dead zones is avoided.

The spiral dryer, being a contact dryer, compares favorably with convective dryers with regard to energy consumption. The use of a relatively small quantity of gas is an additional advantage when operating in a closed circuit. Carbon steel and stainless steel are typically used as materials of construction of the spiral dryer along with special materials.

Since the dryer serves the chemical and the food industries, typical products that are processed include polymer powders, carboxy methyl cellulose, and various light organic and inorganic powders. Furthermore, several flour dryers of this type also exist. Heating, cooling, and thermal treatments are also possible.

The heating medium can be steam, warm water, or a thermal oil. Thus, process temperatures up to approximately 300°C are possible. Steam pressures can reach 20 bar (or 210°C).

The number of kg of product per kg of gas can be greater than that for flash dryers (maximum 1) (i.e., up to 4 and even up to 8 in some cases). Thus, the sizes of the gas-handling equipment are modest. The residence time of the product is in the range of 3–15 sec and, hence, is definitely longer than in a flash dryer. This fact combined with the great driving force for moisture release as far as the exit ensures that, in some instances, materials with bound moisture can be dried satisfactorily.

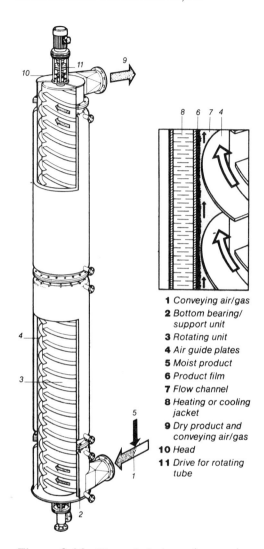

1 *Conveying air/gas*
2 *Bottom bearing/
  support unit*
3 *Rotating unit*
4 *Air guide plates*
5 *Moist product*
6 *Product film*
7 *Flow channel*
8 *Heating or cooling
  jacket*
9 *Dry product and
  conveying air/gas*
10 *Head*
11 *Drive for rotating
  tube*

Figure 9.16 The spiral dryer (seen schematically). (Courtesy of Werner & Pfleiderer Co., Stuttgart, Federal Republic of Germany.)

Werner & Pfleiderer Company (the manufacturer) indicates that heat-transfer coefficients in the range of 90–700 W/m^2 · K can be attained. Aspects having pronounced influences on the actual value include: (a) the nature of the moisture to be removed, (b) the product's properties, and (c) the loading. The pressure loss of the gas on passing through the spiral dryer is in the range of 500–2,000 mm wg.

Figure 9.17 Two spiral dryers that are employed for polyethylene and polypropylene processing. The diameter is 1,600 mm, and the total height is 19 m. (Courtesy of Werner & Pfleiderer Co., Stuttgart, Federal Republic of Germany.)

Pilot plant trials are carried out in a dryer having a diameter of 150 mm and having a length of 5 or 10 m. After taking experience into account, the results are being used for scaling up purposes by the manufacturer. Spiral dryer diameters vary between 150 and 2,000 mm and their lengths between 5 and 30 m (cf. Fig. 9.17).

9.7 Mildly agitated contact dryer (paddle dryer)

Figures 9.18 and 9.19 exhibit the functioning of typical paddle dryers. The wet feed enters at the left and exits at the right. The agitators and the jacket

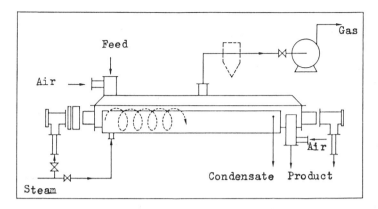

Figure 9.18 The Nara Paddle Dryer (seen schematically). (Courtesy of Goudsche Machinefabriek, Gouda, The Netherlands.)

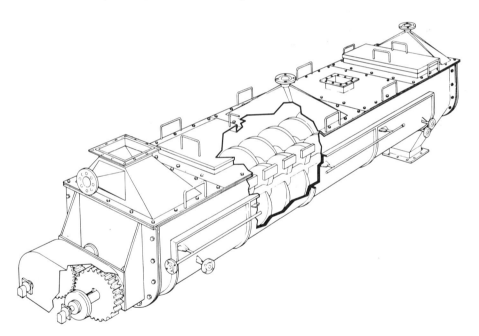

Figure 9.19 A paddle dryer with twin agitators in an omega-shaped trough. (Courtesy of Komline-Sanderson Engineering Corp., Peapack, New Jersey.)

are heated, and the evaporated moisture is entrained by a small amount of carrier gas entering at both ends and sucked off centrally.

The agitators (either 2 or 4) consist of shafts and paddles that mix the mass without exerting any pumping action. Essentially, the mass is fluidized to some degree, which causes the product to flow through the dryer. Evaporation of both water and solvents is possible and heating, cooling, and calcining can be carried out. However, contact of the product with a warm or hot wall cannot be avoided.

Because the amount of entraining gas is relatively small, the gas/solid separation is simple. Because of its construction, the paddle dryer provides ample residence times. The drying of moderately wet, fine, and light materials is a typical application: e.g., gypsum, pigments, dyestuffs, and polyethylene terephthalate. The smallest dryer has a heat-exchanging area of 2.8 m^2; the largest offers 165 m^2. The hold-up volumes range from 65 l to 16.5 m^3. Typical rotational speeds are between 10 and 30 rpm. A maximum steam pressure of 12 bar can be employed. With thermal fluids, temperatures of approximately 300°C can be used.

Typical heat-transfer coefficients vary from approximately 100 to about 550 W/m^2 · K. The higher values are found for products that have surface moisture adhering to them. The lower values are applicable for cooling and heating dry particulate materials. The rotational speed has an influence on the overall heat-transfer coefficient. Clearly, large evaporative loads require large pieces of equipment. The surface/volume ratio of this type of paddle dryer decreases on scaling up. The surface increases proportionally with d^2 and the volume increases proportionally with d^3. Hence, $s/v \sim 1/d$.

If scaling up is carried out on the basis of the area, the residence time increases in proportion with d. Results of pilot plant trials with a 2.8-m^2 dryer can be scaled up. An installed paddle dryer can be seen in Figure 9.20.

Sample calculation

Semitechnical experiments are carried out in a Nara Paddle Dryer, with a contact area of 2.8 m^2. Steam is used as the heating medium, with a condensation temperature of 125°C. The water evaporation occurs at atmospheric pressure. Air is used to entrain the evaporated water, and both gas inlet and outlet temperatures are approximately 100°C. The inlet moisture content (wet basis) is 22.5% by wt and the product's moisture content (wet basis) must be below 0.5% by wt (wet basis).

Heat of evaporation of water at 0°C: 2,504 kJ/kg
Specific heat of water: 4.19 kJ/kg · K
Specific heat of solid: 1.0 kJ/kg · K
Feed temperature: 20°C

Figure 9.20 A paddle dryer processing cellulose acetate wash for recovery purposes. (Courtesy of Komline-Sanderson Engineering Corporation, Peapack, New Jersey.)

The test at which the product's temperature is 100°C throughout and the final moisture content (wet basis) is 0.2% by wt is selected for scaling up. The throughput is 90 kg of feed per hr, feeding 100 kg/hr results in a wet product.

Mass balance, kg/hr

	In	Out
H_2O	20.25	0.14
solids	69.75 +	69.75 +
	90	69.89

Evaporation 20.25 − 0.14 = 20.11 kg/hr

Heat balance, kJ/hr

Process flow heat requirement

$$20.11(2{,}504 + 1.886 \times 100 - 4.19 \times 20) = 52{,}463$$
$$69.75 \times 1.0(100 - 20) = 5{,}580$$
$$0.14 \times 4.19(100 - 20) = 47$$
$$\overline{} +$$
$$58{,}090$$

Heat transfer coefficient, $W/m^2 \cdot K$

$$\frac{58,090 \times 1,000}{3,600 \times 2.8(125 - 100)} = 230.5$$

Scaling up

The desired industrial output is 600 kg/hr.

Dryer area: $2.8 \times \dfrac{600}{69.89} = 24.0 \ m^2$

Take a standard dryer with 28.0-m^2 area.
Jacket steam consumption is (2,200 kJ/kg of steam):

$$58,090 \times \frac{600}{69.89} \times \frac{1.1}{2,200} = 249.3 \text{ kg/hr}$$

Note that a factor of 1.1 is used to allow for heat losses.

Air-heater steam consumption

Industrial evaporation $20.11 \times \dfrac{600}{69.89} = 172.6 \text{ kg/hr}$

Air flow: 1,500 m^3/hr (10°C, 60% RH)
Specific air mass: 1.25 kg/m^3

Kg/hr

	Airflow in	Evaporation	Airflow out
Dry air	1,866.7	0.0	1,866.7
Water vapor	9.3 +	172.6 +	181.9 +
	1,876.0	172.6	2,048.6

Dewpoint exhaust gas is 52°C and the specific air heat is 1.0 kJ/kg · K.
Steam requirement

$$\frac{1.1 \times 1,500 \times 1.25 \times 1.0 \ (100-10)}{2,200} = 84.4 \text{ kg/hr}$$

$249.3 + 84.4 = 333.7$ kg/hr (say, 350 kg/hr) or approximately 2 kg of steam/kg of evaporated water.

9.8 Vigorously agitated contact dryer

A further type of continuous dryer is built as a horizontal jacketed pipe with an agitator. Air travels cocurrently or countercurrently with the prod-

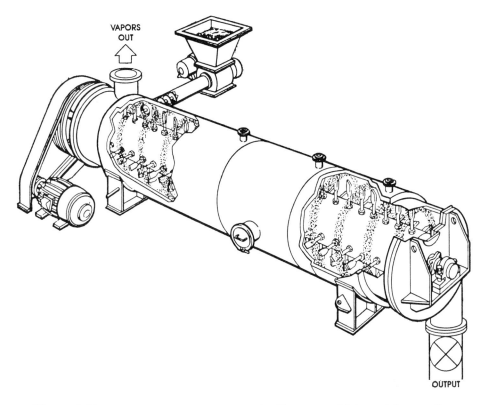

Figure 9.21   Solidaire (seen schematically). (Courtesy of Bépex, Minneapolis, Minnesota.)

uct through this contact dryer (see Fig. 9.21). The heat is supplied by means of the jacket and the air; however, the jacket contribution is usually more important than the convective contribution. The product travels along the wall from the inlet to the outlet. A criterion to be observed here is the Froude number (Fr) which gives the ratio between the radial acceleration and the acceleration that is due to gravity. The radial acceleration is v^2/R, with v being the circumferential velocity and R the radius. The acceleration that is due to gravity is g m/sec^2.

$$Fr = \frac{v^2}{R \times g}$$

If $Fr > 1$, then the product moves in a circular path along the wall. The heat-transfer coefficient can hardly be increased in this region by rotating faster. For all Froude numbers exceeding 1, the heat transfer coefficient is dependent on: (a) the mass fed per unit of time and per m circumference, (b) the moisture content and the material properties, and (c) the paddle position in combination with the rotational speed. The specific dryer load and the paddle position influence the degree to which the heat-transfer area is utilized.

The dryer is particularly suitable for processing cakes that contain organic solid material. Feeds that exhibit a tendency to cake in flash dryers can often be processed in a vigorously agitated contact dryer. The wetter the feed is, generally the less suitable is this type of dryer. The vigorous agitation may lead to a reduction of particle size of material. In addition, the dryer is not suitable for abrasive materials. A typical dryer has hundreds of paddles moving with a tip speed between 10 and 20 m/sec. Residence times between 2 and 20 minutes are possible. Shorter residence times are registered on operating cocurrently, and longer times are obtained for countercurrent operation. The residence time is also influenced by the position of the rotor elements.

An option is to combine a vigorously agitated contact dryer with a dryer providing residence time. Such a combination is used in the drying of polyolefins. Maximum heating-medium, and drying-air temperatures are around 200°C. Typical gas-exit velocities are in the range of 2–4 m/sec.

Figure 9.22 A vigorously agitated contact dryer in an Akzo Chemicals plant (area 3.8 m^2, stirrer speed 500 rpm.)

Typical heat-transfer coefficients vary from about 100 to approximately 400 W/m^2 · K (e.g., 100 for dry PVC powder and 200–250 for a normal drying problem).

The smallest Solidaire dryer (from Bépex) has a diameter of 8 in and a barrel length of 4 ft. This dryer is also used for testing. Further testing can be done on a 16 in by 10 ft dryer. The largest dryer is an 88 in by 50 ft dryer (cf. Fig. 9.22).

Worked example

The problem is to dry a belt filter cake. The solid material is an organic chemical. Preliminary tests exhibited brown incrustations on applying a jacket temperature of 180°C. A jacket temperature of 160°C (6-bar saturated steam) was satisfactory.

Initial moisture content (wet basis): 20% by wt
Final moisture content (wet basis): 1% by wt
Desired large-scale output: 750 kg/hr
Solid specific heat: 1.25 kJ/kg · K

Semi-technical dryer data
Diameter: 0.25 m
Drying cylinder length: 2.3 m
Jacket heat-exchange area: 1.5 m^2

Rotational speed is continuously adjustable from 0 to 1,300 rpm. The following data apply for the experiment selected for scaling up.

Dry product: 150 kg/hr
Product inlet/outlet temperatures: 20/75°C
600 m^3/hr of air at 20°C (relative humidity 60%)
Air inlet/outlet temperatures: 170/100°C
Initial/final moisture content (wet basis): 20/0.5% by wt
Agitator rotational speed: 1,300 rpm
Jacket temperature: 160°C

Interpretation of experimental data
Mass balance, kg/hr

	In	Out
H$_2$O	37.30	0.75
solid	149.25 +	149.25 +
	186.55	150

Evaporation 37.30 − 0.75 = 36.55 kg/hr

Net heat, kJ/hr

$$Q_1 = 36.55(2{,}504 + 1.886 \times 100 - 4.19 \times 20) = 95{,}352$$
$$Q_2 = 149.25 \times 1.25(75 - 20) \qquad\qquad = 10{,}261$$
$$Q_3 = \quad 0.75 \times 4.19(75 - 20) \qquad\qquad\quad = \underline{\quad 173}$$
$$\qquad\qquad\qquad\qquad\qquad\qquad\qquad\qquad\qquad 105{,}786$$

Air contribution

$600 \times 1.2 \times 1.0(170 - 100) = 50{,}400$ kJ/hr
Spent on the process: approximately $0.7 \times 50{,}400 = 35{,}280$ kJ/hr
$0.3 \times 50{,}400 = 15{,}120$ kJ/hr is lost

Jacket contribution

$105{,}786 - 35{,}280 = 70{,}506$ kJ/hr

$$(\Delta T)_m = \frac{(160 - 20) - (160 - 75)}{log_e \dfrac{(160 - 20)}{(160 - 75)}} = 110.2 \text{ K}$$

$$U = \frac{70{,}506 \times 1{,}000}{3{,}600 \times 1.5 \times 110.2} = 118.5 \text{ W/m}^2 \cdot \text{K}$$

The method of calculating U by using the logarithmic mean temperature difference can be criticized.

Agitator tip speed and Fr

$$v = \frac{\pi \times 0.25 \times 1{,}300}{60} = 17.0 \text{ m/sec}$$

$$Fr = \frac{17.0^2 \times 2}{0.25 \times 9.81} = 236$$

Scaling up

Mass balance, kg/hr

	In	Out
H_2O	185.6	7.5
solid	742.5	742.5
	928.1	750.0

Evaporation $185.6 - 7.5 = 178.1$

Net heat, kJ/hr   Air inlet/exit temperatures are, respectively, 170 and 100°C. Product inlet/exit temperatures are, respectively, 20 and 75°C.

$$Q_1 = 178.1(2{,}504 + 1.886 \times 100 - 4.19 \times 20) = 464{,}627$$
$$Q_2 = 742.5 \times 1.25(75 - 20) \qquad\qquad\quad = 51{,}047$$
$$Q_3 = 7.5 \times 4.19(75 - 20) \qquad\qquad\quad\;\; = 1{,}728$$
$$\overline{\qquad\qquad\qquad\qquad\qquad\qquad 517{,}402}$$

Air contribution   A dryer having a diameter of 0.6 m is considered. An air flow of 3,000 m³/hr at 20°C is taken. Heat transferred to the process flow: $0.7 \times 3{,}000 \times 1.2 \times 1.0(170 - 100) = 176{,}400$ kJ/hr.

Jacket contribution $517{,}402 - 176{,}400 = 341{,}002$ kJ/hr
Check of the heat-transfer coefficient

$$(\Delta T)_m = 110.2 \text{ K}$$

Jacket area of the 0.6-m dryer: 6.6 m²

$$U = \frac{341{,}002 \times 1{,}000}{3{,}600 \times 110.2 \times 6.6} = 130.2 \text{ W/m}^2 \cdot \text{K}$$

This is a higher value than found by experiment. However, the specific load is approximately 2 times larger now, and it is possible to adjust the paddle position.

Agitator tip speed and Fr   Agitator rotational speed: 600 rpm

$$v = \frac{\pi \times 0.6 \times 600}{60} = 18.8 \text{ m/sec}$$

$$Fr = \frac{18.8^2 \times 2}{0.6 \times 9.81} = 120$$

A dryer having a diameter of 0.6 m is a good choice.

9.9   Vertical thin-film dryer

A vertical thin-film dryer is a continuous agitated contact dryer that can convert a suspension or a solution into a dry powder (see Fig. 9.23).

This equipment originated from the thin-film evaporator that was used to concentrate solutions. The design is sturdy and can process pastes and powders (see Fig. 24). The power consumed during the processing of the paste—the intermediate stage between fluid and solid—can be quite high.

The fluid is fed at the top and flows down. The agitator rotates at speeds corresponding with *Fr* numbers that exceed unity, which means that the

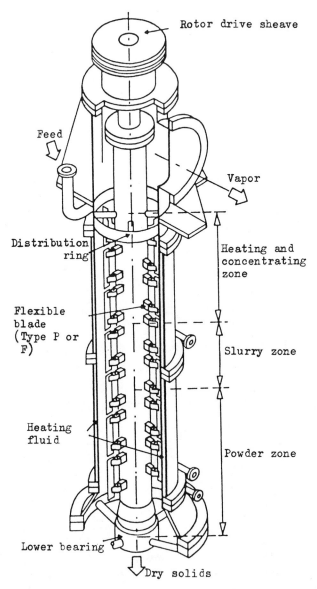

Figure 9.23   A vertical thin-film dryer. (Courtesy of Buss, Basel, Switzerland.)

Figure 9.24 A vertical thin-film dryer equipped with vacuum locks for powder extraction. (Courtesy of Buss, Basel, Switzerland.)

product moves in a spiral-shaped path along the wall. The agitator elements are not rigid but have a hinge halfway between the shaft and the wall. Thus, a wall-scraping action is exerted that is in itself safeguarded against damage. The clearance between the wall and the elements is 0.3–0.5 mm. The drying process can be carried out at atmospheric pressure or under vacuum. Typical applications are: (a) recovering solvents, (b) crystallizing and drying of salts from solutions, and (c) obtaining pigments from aqueous slurries.

Normally, the pressure in the dryer can be chosen from 1 Torr up to 1 bar. Special constructions are adequate for higher pressures. The maximum temperature is 195°C when using steam and 350°C when using a thermal oil. Circumferential rotor velocities are in the range of 6–8 m/sec.

Figure 9.25 A combination of a vertical thin-film dryer and a vigorously agitated horizontal contact dryer. (Courtesy of Buss, Basel, Switzerland.)

Typical residence times vary from 10–20 sec per meter dryer length. Normal feed rates can be recalculated to a figure in the range of 0.8–1.2 kg of feed per cm circumference per hr. Because of the relatively short residence time, it is not too easy to obtain final moisture contents lower than 0.1% by wt.

Semi-technical test results can be scaled up by a maximum factor of 50. Kg of feed processed per hr and by m^2 heat transfer area is the basis. Commercial dryer sizes are in the range of 0.5–24 m^2.

If it is necessary to reach final moisture contents as low as 0.01% by wt, it could be advantageous to combine the vertical thin-film dryer with the vigorously agitated contact dryer (see Fig. 9.25). The residence time can be stretched up to 10 min per m in the horizontal dryer length. Typically, the horizontal dryer's process variables are numerically comparable to the process variables mentioned for the vertical thin-film dryer. The circumferential velocities are usually in the range of 8–10 m/sec.

When it is necessary, for safety reasons or for hygienic reasons, to combine several unit operations in one piece of equipment, the combination of the two dryer types is attractive.

9.10   Steam-tube dryers

Indirect rotary dryers can be subdivided into low- and high-temperature devices. The former are usually heated by steam, the latter are fired either by oil or natural gas.

The steam-tube dryer is one of the most common types of indirect rotary dryers. It is a contact dryer in the form of a slowly rotating almost horizontal cylinder, with heat-transfer tubes along the circumference. The material to be dried enters at one end and exits at the other end. The shell is inclined at a slight angle (1°–5°). Steam is introduced into the tubes through a manifold located at the discharge end of the cylinder. The manifold is connected to a rotary steam joint that admits the steam and continuously drains the condensate.

There is a natural-draft stack to remove air and vapors. The purge passes through a cyclone, a settling chamber, or a wet scrubber to separate any entrained solids. A steam-tube dryer's central discharge assembly is depicted in Figure 9.26. The dryer is useful for fine dusty particles that would be entrained by the gas in a convective dryer. Usually, there is a slight underpressure in the dryer.

The steam pressure is normally in the range of 4–10 bar; however, the pressure may be as high as 40 bar. The steam tubes together with the radial flights serve to agitate the contents in the dryer. As many as three staggered, concentric rows of tubes are possible. The diameter of the tubes of

Figure 9.26 A central discharge assembly on an 8 by 60 ft rotary steam-tube dryer. (Courtesy of Swenson Inc., Harvey, Illinois.)

the inner rows is smaller than the diameter of the tubes of the outermost row. The pitch-to-diameter ratio of the tubes must be sufficiently large (e.g., 2:1) to ensure that the solid material can flow freely around them. For sticky materials, normally a single row of tubes is used only. Backmixing and shell rappers may also be used.

The rotating shell is sealed into the two end breechings. Various constructions exist. The length-to-diameter ratio can vary from 4:1 to 10:1. The circumferential velocity is normally in the range of 0.3–1 m/sec. Percentages by volume of material vary from 10 to 20. The heat-transfer coefficients are in the range of 30–90 W/m² · K.

Semitechnical experiments are, for example, carried out in a dryer having a 12-in diameter and a length of 8 ft. It is possible to have 8 tubes with a diameter of 1 in.

Prices of steam-tube rotary dryers are listed in Table 9.1.

Table 9.1 Steam-Tube Dryers[a,b]

Diameter (m)	Length (m)	Transfer area (m²)	Rotation speed (rpm)	Size (kW)	Price (as of 1986)
.965	4.572	21.5	6	2.2	$145,000
.965	6.096	29.1	6	2.2	160,000
.965	7.620	36.8	6	3.7	170,000
.965	9.144	44.4	6	3.7	180,000
.965	10.668	52.1	6	3.7	190,000
1.372	6.096	58.3	5	3.7	185,000
1.372	7.620	73.5	5	3.7	200,000
1.372	9.144	88.8	5	5.6	215,000
1.372	10.668	104.1	5	5.6	230,000
1.829	7.620	118.2	4	5.6	225,000
1.829	9.144	142.8	4	5.6	240,000
1.829	10.668	168.0	4	7.5	255,000
1.829	12.192	192.0	4	7.5	270,000
1.829	13.716	216.6	4	11.2	285,000
1.829	15.240	241.3	4	11.2	300,000
1.829	16.764	265.9	4	14.9	315,000
1.829	18.288	290.5	4	14.9	330,000
2.438	12.192	384.2	3	11.2	420,000
2.438	13.716	443.2	3	11.2	470,000
2.438	15.240	482.6	3	14.9	510,000
2.438	16.764	531.9	3	14.9	545,000
2.438	18.288	581.1	3	14.9	580,000
2.438	19.812	630.3	3	18.7	620,000
2.438	21.336	679.6	3	22.4	660,000
2.438	22.860	728.8	3	25.9	695,000
2.438	24.384	778.1	3	29.8	730,000
3.200	13.716	650.0	2.25	14.9	620,000
3.200	15.240	723.9	2.25	18.7	660,000
3.200	16.764	797.9	2.25	22.4	710,000
3.200	18.288	871.7	2.25	25.9	750,000
3.200	19.812	945.6	2.25	29.8	790,000
3.200	21.336	1019.4	2.25	33.5	825,000
3.658	15.240	1000.8	1.5	22.4	800,000
3.658	16.764	1102.9	1.5	25.9	850,000
3.658	18.288	1205.1	1.5	29.8	900,000
3.658	19.812	1307.2	1.5	33.5	945,000
3.658	21.336	1409.3	1.5	37.3	990,000
3.658	22.860	1511.4	1.5	41.0	1,040,000
3.658	24.384	1613.6	1.5	44.7	1,085,000
3.658	25.908	1715.7	1.5	48.4	1,130,000
3.658	27.432	1817.8	1.5	52.2	1,175,000

Table 9.1 (Continued)

Diameter (m)	# Rows	#1 # (Dia.)	#2 # (Dia.)	#3 # (Dia.)	#4 # (Dia.)
.305	1	8 (25)			
.610	1	12 (60)			
.965	1	14 (114)			
1.372	2	18 (114)	18 (64)		
1.829	2	27 (114)	27 (76)		
2.438	3	24 (114)	30 (114)	36 (114)	
3.200	3	45 (114)	45 (114)	45 (114)	
3.658	4	54 (114)	54 (114)	54 (76)	54 (76)

[a]Carbon steel fabrication; multiply prices by 1.7 for 304 stainless steel. (Courtesy of Swenson Inc., Harvey, Illinois.)
[b]Tube diameters in mm.

Sample calculation

Very fine material must be dried. The filter cake contains 75% by wt of water (wet basis). Tests are carried out in a small-scale steam-tube dryer. It is possible to obtain a material containing less than 1% water by wt. The exit temperature of the material is 120°C. Processing capacity is 2.0 kg of product per hr.

Dryer

Diameter: 12 in
Length: 8 ft
Number of tubes: 8
Heat transfer area: 1.56 m²(16.75 ft²)

Steam

Pressure: 8 bar
Temperature (saturated): 169°C
Latent heat: 2,051 kJ/kg

Calculation

Mass balance, kg/hr

	In	Out
solids	1.98	1.98
water	5.94 +	0.02 +
	7.92	2.0

Evaporation $5.94 - 0.02 = 5.92$ kg/hr

Heat balance, kJ/hr

$$Q_1 = 5.92\,(2{,}504 + 1.886 \times 100 - 4.19 \times 40) = 14{,}948.0$$
$$Q_2 = 1.98 \times 0.75\,(120 - 40) \qquad\qquad = \quad 118.8$$
$$Q_3 = 0.02 \times 4.19\,(120 - 40) \qquad\qquad = \quad\quad 6.7 \;+$$
$$\overline{15{,}073.5}$$

Heat-transfer coefficient   Assume that the heat flow conducted through the tube wall is $1.2 \times 15{,}073.5 = 18{,}088.2$ kJ/hr (heat losses and air heating-up).

$$U = \frac{18{,}088.2 \times 1{,}000}{(169 - 100)1.56 \times 3{,}600} = 46.7 \text{ W/m}^2 \cdot \text{K}$$

Commercial dryer capacity: 1,000 kg/hr

Mass balance, kg/hr

	In	Out
solids	990	990
water	2,970 +	10 +
	3,960	1,000

Evaporation 2,960 kg/hr

Heat balance, kJ/hr

$$Q_1 = 2{,}960\,(2{,}504 + 1.886 \times 100 - 4.19 \times 20) = 7{,}722{,}048$$
$$Q_2 = 990 \times 0.75\,(120 - 40) \qquad\qquad = \quad 59{,}400$$
$$Q_3 = 10 \times 4.19\,(120 - 40) \qquad\qquad = \quad\quad 3{,}352$$
$$\overline{7{,}784{,}800}$$

Multiply by 1.1 to allow for air heating up and heat losses: $1.1 \times 7{,}784{,}800 = 8{,}563{,}280$ kJ/hr

Dryer size   Required heat-transfer area

$$\frac{8{,}563{,}280 \times 1{,}000}{(169 - 100)46.7 \times 3{,}600} = 738.2 \text{ m}^2$$

From the data in Table 9.1:

Dia (m) × Length (m) = 2.438 × 24.384
Price: $730,000 (carbon steel)

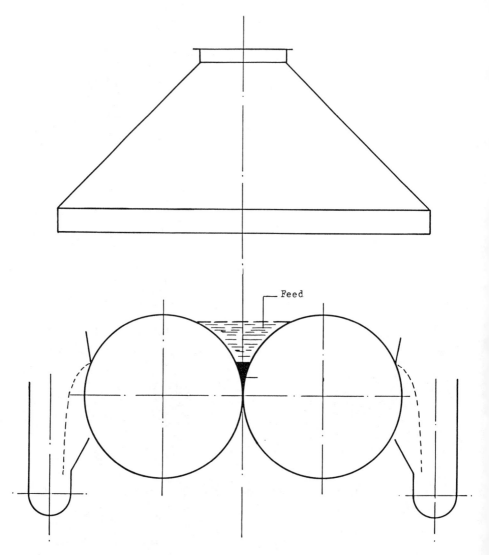

Figure 9.27 A double-drum dryer for low-viscosity solutions. (Courtesy of Goudsche Machinefabriek, Gouda, The Netherlands.)

9.11 Drum dryers

A drum dryer processes materials effecting a thermal solid/liquid separation by applying the feed onto one or several slowly rotating steam-heated drums and removing the product by means of a knife. Solutions, suspensions, and pastes can be fed, and powders, flakes, or chips can be obtained. Evaporated water is extracted by means of a vapor hood. It is also possible to evaporate solvents and, in order to keep the temperature down, to operate under a vacuum.

The drum dryer was introduced into the processing industry approximately one hundred years ago. Development started with the double-drum dryer, featuring the feed flowing into the nip between the drums (see Figs. 9.27 and 9.28). The product layer's thickness is a function of the distance between the drums and can be selected from the range of 100 to 400 microns.

The double-drum dryer is still being used widely; however, because pressure on the rolls is considerable, the dryer is less suitable for viscous feedstocks. This fact led to the introduction (in 1945) of the top-feed single-drum dryer (see Figs. 9.29 and 9.30) which can handle viscous feedstocks. Furthermore, for special applications, it is possible to apply the feed by means of dipping, splashing, spraying, or a bottom-feed roll.

Drum dryers are being used in those instances where conventional methods to separate solids and liquids (e.g., filtration and convective drying) might fail. Typical materials dried on a double-drum dryer include baby food, devitalized yeast, glues, and dairy products. On a single-drum type, starch, instant potatoes, dyestuffs, and gelatin.

Figure 9.28 An installed double-drum dryer. (Courtesy of Goudsche Machinefabriek, Gouda, The Netherlands.)

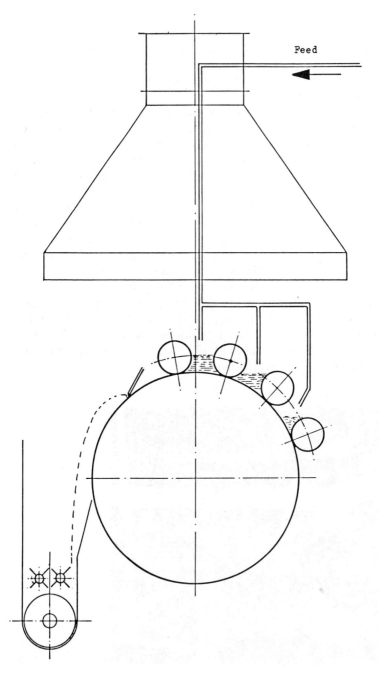

Figure 9.29 A single-drum dryer with applicator rolls for the processing of pastes and highly viscous products. (Courtesy of Goudsche Machinefabriek, Gouda, The Netherlands.)

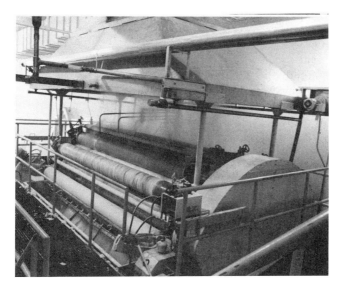

Figure 9.30 An installed single-drum dryer. (Courtesy of Goudsche Machine-fabriek, Gouda, The Netherlands.)

The drum-drying system has limitations. Specifically, materials that do not adhere to metal cannot be processed, materials that cannot stand the temperature (considering the exposure time) are unsuitable, and materials that are or get too viscous cannot be handled.

Drum dryers can be built up to a diameter of 2 m and a length of 5 m. Usually, the capacities are in the range 5–30 kg of product / m^2 · hr. The maximum capacity of the single-drum dryer is thus approximately 1 t/hr. Residence times vary from 2 sec to 1 min. The maximum steam pressure normally is 12 bar and the rotational speeds are in the range 1–30 rpm.

Temperature-sensitive materials like concentrated milk and gelatin are processed at high speeds. The drums can be made out of cast iron or welded steel. Chrome hardening of the drum surface is practiced because of corrosive feedstocks.

Semitechnical experiments are typically carried out on a single- or double-drum dryer, the drums having both a diameter and a length of 0.5 m. Generally speaking, scaling up is based on the mass flows per unit area.

A suitable temperature and a feasible feeding arrangement is selected. The residence time is fixed by means of the rotational speed.

Pricing of drum dryers varies from $50,000 for the smallest dryer up to $400,000 for the largest one (according to data supplied [in 1987] by Goudsche Machinefabriek, Gouda, The Netherlands).

Notation

d	Dryer dimension	m
Fr	Froude number	
g	Acceleration due to gravity	m/sec^2
Q_1	Heat flow to evaporated water	kJ/hr
Q_2	Heat flow for the dry solids	kJ/hr
Q_3	Heat flow for the residual water	kJ/hr
R	Radius of dryer	m
s	Heat-transfer area in dryer	m^2
U	Heat-transfer coefficient	W/m$^2 \cdot$ K
v	Circumferential velocity	m/sec
	Dryer volume	m^3
$(\Delta T)_m$	Logarithmic mean-temperature difference	K
ρ	Specific mass	kg/m^3

10

BATCH DRYERS

10.1 Introduction

Both convective and contact dryers are discussed in this chapter. Fluid-bed and atmospheric tray dryers are examples of convective dryers; the fluid-bed dryer is the more powerful one. The atmospheric tray dryer is limited by relatively high personnel costs because of the amount of physical handling required in its operation.

Contact dryers offer the advantage of containment, and the vacuum types offer the additional bonus of low-temperature drying. The agitated vacuum dryer, the tumbler, the vacuum-tray dryer, and the agitated atmospheric dryer are covered in this chapter. Also, examples of scale-up calculations starting from test work are presented. Batch dryers tend to be used for organic speciality chemicals and pharmaceuticals.

10.2 Batch fluid-bed dryers

A batch fluid-bed dryer processes wet particulate material that can be dried by blowing hot or warm air through its mass. There are similarities with the continuous fluid-bed dryers that were described in Chapter 5. The material to be dried is charged in a layer between 0.3 and 1 m thick into a container with a perforated supporting plate. The container is placed in a drying cabinet and airflow through the layer causes drying. Gradually, as the material gets drier, fluidization starts. The air leaves the dryer through filter bags (see Fig. 10.1 and also Fig. 3.4 in Chapter 3).

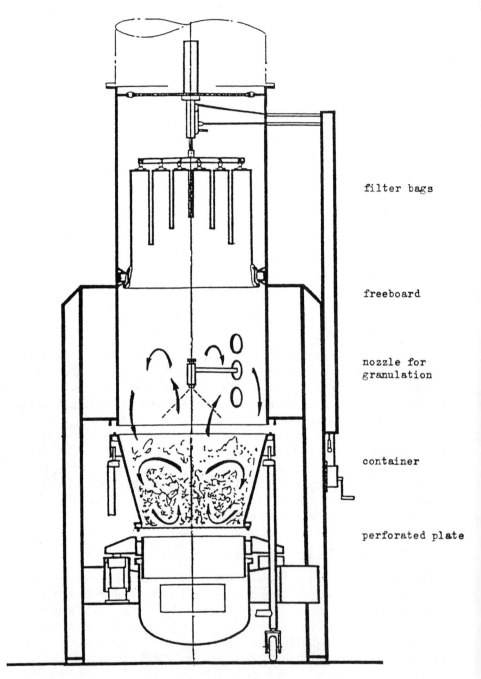

filter bags

freeboard

nozzle for
granulation

container

perforated plate

Figure 10.1 A batch fluid-bed dryer with a nozzle for granulation. (Courtesy of Glatt GmbH, Binzen, Federal Republic of Germany)

The air velocities can be quite high in these dryers and proper fluidization may not be achieved. This is not too much of a problem since the material cannot be carried over. It is a common practice to blow air through the bed for a while and then interrupt the gas flow to allow material on the filter cloth to fall down. This type of dryer is widely used for drying fine chemicals and pharmaceuticals. The evaporation of free moisture can be described with the aid of the Mollier H/y-diagram. The basic design equation is:

$$G_{H_2O} = G_{air}(y_o - y_i) \text{ kg} \tag{10.1}$$

Chapter 5 describes cases where the simple model was found to be applicable to drying from the initial moisture content down to the critical moisture content. Usually, at the end of the first drying step, the material still contains too much moisture. A reduction of the moisture content from a few tenths of a percent to almost nil is often still required. As the evaporation duty is usually small, the second drying step can be described as heating up the granular material whereby thermal equilibrium is reached (i.e., the temperature of the fluidized bed equals that of the exit gas). The rate of temperature rise is proportional to the driving force $[TA_{IN} - T_p(t)]$.

$$\frac{dT_p(t)}{dt} = \frac{G_{gas} \cdot C_p[TA_{IN} - T_p(t)]}{G_s \cdot CP}$$

Integration gives:

$$t_2 = \frac{G_s \cdot CP}{G_{gas} \cdot C_p} \cdot \log_e \left[\frac{TA_{IN} - T_p \, (t=o)}{TA_{IN} - T_p \, (t=t_2)} \right] \text{ sec} \tag{10.2}$$

However, one may also encounter substances where the situation is more complicated and the need to evaporate a considerable amount of bound moisture exists. For example, this can be caused by the granule structure when there are crevices and/or pores. In cases of diffusion limited evaporation, the temperature gradually increases and the moisture content gradually decreases. The time requirements must be established experimentally. See Chapter 5 for information concerning the applicability of these two design equations.

Cooling is carried out prior to packaging and can be dealt with in the same way as the evaporation of bound water. Batch fluid-bed dryers may contain up to 1.5 m³ of material. The corresponding fan capacity is of the order of 12,000 m³/hr. Drying-gas temperatures seldom exceed 200°C and drying with an inert gas (usually nitrogen) is also possible. Small-scale trials can be carried out in dryers containing up to 45 l of material. It is possible to combine batch fluid-bed drying with fluid-bed granulation and fluid-bed coating.

Sample calculation

Calculate the cycle time for a fluid-bed drying operation.

Feed

Batch size, kg (wet): 120
Initial moisture content, % by wt (wet basis): 30
Final moisture content, % by wt: 0.0
Critical moisture content, % of wt: 0.1
Specific heat solid material, J/kg · K: 1,200
Temperature, °C: 18

Drying gas/cooling gas

Air is being taken from the atmosphere. The conditions are: 10°C and 50% relative humidity. The air is warmed up to 40°C in the drying period. Gas velocity under the distribution plate is 0.7 m/sec:

Dryer

Diameter, m: 0.9
Pressure under the distribution plate: atmospheric
The air is warmed indirectly with warm water.

Calculation

First drying period

Equation (12.1): $G_{H_2O} = G_{air} \times \Delta y$ kg

Mass balance product, kg

	In	Out
solid	84	84
H_2O	$\dfrac{36}{120}+$	$\dfrac{0.1}{84.1}+$

Evaporation $36 - 0.1 = 35.9$ kg

$\Delta y = 0.013 - 0.004 = 0.009$ kg H_2O/kg dry air

See Mollier's diagram.

$$G_{air} = \frac{35.9}{0.009} = 3{,}988.9 \text{ kg}$$

Note: this is dry air.

Warm airflow (40°C)

	kg	kg/m^3	m^3
dry air	3,988.9	1.13	3,530
H_2O	$\dfrac{16.0}{4,004.9}+$	0.70	$\dfrac{23}{3,553}+$

$$t_1 = \frac{3,553}{\pi/4 \times 0.9^2 \times 0.7} = 7,979 \text{ sec } (2 \text{ hr and } 13 \text{ min})$$

Second drying period

$$C_p = \frac{3,988.9}{4,004.9} \times 1,000 + \frac{16.0}{4,004.9} \times 1,886 = 1,004 \text{ J/kg} \cdot \text{K}$$

Equation (5.2): $t_2 = \dfrac{G_s \cdot CP}{G_{gas} \cdot C_p} \cdot \log_e \left[\dfrac{TA_{IN} - T_p\,(t=o)}{TA_{IN} - T_p\,(t=t_2)} \right]$ sec

Warming up to 35°C is required to get a completely dry material. The adiabatic saturation temperature $[T_p(t = o)]$ is 18°C.

$$t_2 = \frac{84 \times 1,200}{\pi/4 \times 0.9^2 \times 0.7 \times 1,004 \times 1.13} \log_e \left(\frac{40 - 18}{40 - 35} \right) = 296 \text{ sec } (4.9 \text{ min})$$

Cooling period   Cooling to 20°C prior to packing is required.

$$t_3 = \frac{84 \times 1,200}{\pi/4 \times 0.9^2 \times 0.7 \times 1,004 \times 1.25} \log_e \left(\frac{35 - 10}{35 - 20} \right) = 92 \text{ sec } (1.5 \text{ min})$$

Loading and unloading: 15 min
Spare time: 15 min

Summary

First drying period: 2 hr 13 min
Second drying period: 4.9 min
Cooling period: 1.5 min
Loading / unloading: 15 min
Spare time: 15 min
   Total 2 hr 49.4 min. Say, 3 hr.

10.3   Atmospheric tray dryer

An atmospheric tray dryer is a convective dryer shaped as an enclosed, insulated housing in which trays containing the feed are placed (see Fig. 10.2). The material to be dried is placed in layers of between 1–5 cm thick on trays that are then loaded on trucks which are wheeled into dryer

Figure 10.2   An atmospheric tray dryer. (Courtesy of Proctor & Schwartz, Inc., Horsham, Pennsylvania.)

cabinets. After closing the doors, the air is heated and circulated through the dryer. There is a purge of warm air containing the evaporated moisture and fresh make-up air is introduced. The loading and unloading of the trays require considerable manual labor and although automation is, in principle, possible, this is one aspect that has caused in many instances the replacement of tray dryers with other systems (e.g., fluid-bed dryers).

A further aspect to be considered is the relatively low output of tray dryers when measured in kg/hr · m³. A drying cycle of between 5 and 40 hr is not unusual in tray dryers. Fluid-bed dryers are much superior in this respect. The final drawback is the nonuniform airflow in tray dryers that can result in local overheating even though in other parts of the dryer wet product can still be found. Because of product quality or process safety requirements often a restriction is placed on the drying-air temperature. Large individual items can be stacked in piles or placed on shelves, which demonstrates that the tray dryer can accept a wide range of feed physical forms (e.g., filter cakes, pastes, sludges, etc.). Sometimes, however, the product must be converted into a free-flowing powder by means of a screen or other device.

The most common practice is to circulate the gas between the trays (i.e., the dryers operate with cross-airflow). An alternative is to circulate the air through the material. In the case of coarse particulate material, this leads to higher specific evaporation rates in kg per m^2 tray area per hr. Atmospheric tray dryers are used for the drying of pharmaceuticals, pigments, and fine chemicals.

Normally, a maximum air temperature of approximately 200°C is employed, with the air being indirectly heated by means of steam. An average recirculation figure of 80% is used. Cross-airflow velocities can be as high as 3 m/sec, but a value of 1–1.5 m/sec is customary. Through-circulation equipment typically uses an air velocity of 0.5–1 m/sec.

Typical performance data vary between 0.1 to 1.5 kg of water evaporated per $m^2 \cdot$ hr, with the higher values being obtained at high drying temperature and vice versa (e.g., 0.2 at 50°C and 1.0 at 150°C). Usually, steam consumption varies between 1.5 to 3.5 kg of steam per kg of evaporated water. Normally, tray dryers are from several cubic meters to several tens of cubic meters.

The results of small-scale drying trials can be readily scaled-up. Cooling the product prior to packaging can be easily achieved by maintaining the circulation while removing the heating.

Sample calculation

A pharmaceutical product is to be dried with air of 95°C and subsequently cooled to 40–50°C. A tray dryer with 29 trays is available, area per tray 1 m^2. Approximate tray dryer dimensions: (a) width 1 m, (b) depth 1.5 m, and (c) height 2 m. Initial moisture content (wet basis): 32 % by wt. Bulk density wet product: 600 kg/m^3.

Capacity estimate

Take a layer thickness of 2 cm. Feed charge $29 \times 1 \times 0.02 \times 600 = 348$ kg.

Mass balance product, kg/batch

	In	Out
solid	237	237
H_2O	$\dfrac{111}{348}+$	$\dfrac{0}{237}+$

Evaporation 111 kg/batch

A specific capacity of 0.6 kg/hr $\cdot$ m^2 is assumed.

Drying cycle $\dfrac{111}{29 \cdot 0.6}$ = 6.4 hr

Capacity $\dfrac{237}{6.4}$ = 37.0 kg/hr

10.4 Agitated vacuum dryer

In an agitated vacuum dryer, a batch of wet particulate material is dried under vacuum by heating the agitated material and causing the evaporation of the water. Drying occurs at a reasonable rate while the product temperature is low. Because the process is contained, the agitated vacuum dryer is particularly suitable for the evaporation of solvents or the handling of toxic materials. Stein (1976 and 1977) compared the performance of different types of agitated vacuum dryers and concluded that the cone dryer with screw has good characteristics if all aspects are taken into account. The dryer feed is a filter cake having a tendency to incrustations but ultimately becoming a free-flowing powder. Drying the product from 60% to 0.1% by weight of water occurs with a heat-transfer coefficient of 60 kcal /hr $\cdot$ m^2 $\cdot$ K (70 W/m^2 $\cdot$ K). However, several simplifying assumptions were made. The published data on heat-transfer coefficients exhibit a large scatter (see, e.g., Entrop and Jensen, 1975, and Lücke, 1976). This research underlines the necessity to rely on test results.

Besides the aforementioned good heat-transfer characteristics, the cone dryer with screw has several additional advantages when compared with other types of agitated vacuum dryers:

1. A top-driven stirrer without bottom bearings precludes the penetration of the material being dried into a bearing or packing.
2. The largest diameter is at the top and this keeps the gas velocity low thus minimizing powder entrainment.
3. The head space serves as a freeboard allowing the settling of particles.

Figure 10.3 illustrates a cone dryer with screw. (Chapter 3 contains a photograph of a similar piece of equipment.) The agitator is an Archimedian screw rotating around its own axis and pumping upward and moving over the cone's area. The screw can pump down when discharging the equipment. The evaporation of free moisture occurs while the product temperature is constant and, ideally, this temperature is the boiling point at the prevailing pressure. The resistance to heat transfer is located in the heat flow from the metal wall to the wet particles. Because this resistance is high, the water temperature drop on flowing through the jacket is usually 1–2 K. Scaling-up proceeds by means of the multiplication of the drying

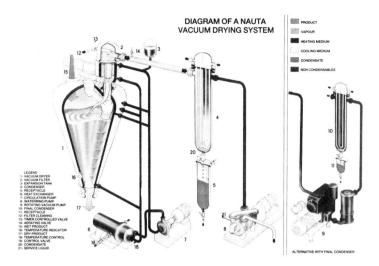

Figure 10.3 A diagram of a Nauta Vacuum Drying System. (Courtesy of Hosokawa-Nauta BV, Haarlem, The Netherlands.)

time by the batch volume to the warmed area ratio, i.e., multiplying the drying time with the cube root of the volumes ratio.

It is then assumed that (a) the degree of filling on a volume basis is equal for the two scales, (b) there is a geometric similarity, (c) the warm water temperatures are equal, (d) the vacuum conditions are equal, and (e) the mixing conditions are equal. If the amount of bound moisture is small, the thermal process in the second drying period can be described as a heating-up of the granular material.

$$U \cdot A[T_i - T_p(t)]dt = G_s \cdot CP \cdot dT_p(t)$$

On integrating and rearranging:

$$U = \frac{G_s \cdot CP}{t_2 \cdot A} \cdot \log_e \left[\frac{T_i - T_p \ (t=0)}{T_i - T_p \ (t=t_2)} \right] \text{ W/m}^2 \cdot \text{K} \qquad (10.3)$$

Thus, the U-value can be found for scaling-up calculations. If the amount of bound moisture cannot be neglected, the thermal process in the second drying period is more complicated however. In this case, the drying time can be multiplied by the cube root of the volumes ratio.

Hosokawa-Nauta BV kindly provided the following information:

1. The standard series with bottom bearing range from 1 to 4 m^3; however, dryers with bottom bearings with volumes of up to 25 m^3 have been built. Note that these figures are given for total volumes.
2. The standard series with top-suspended screws range from 50 l to 2.5 m^3. Nonstandard equipment has been constructed up to 5 m^3.
3. The maximum working temperature is 250°C, whereas maximum and minimum working pressures are 10 bar and 1 mbar, respectively.
4. Test work is carried out in a cone dryer having an effective volume of 300 l.

Sample calculation

Predict the capacity of a 1,000–l cone dryer with screw based on the performance data from a 300-l dryer (effective volumes are quoted).

Testing on a 300-l dryer
Feed

Batch size, kg (wet): 150
Initial moisture content, % by wt (wet basis): 8.0
Final moisture content, % by wt: 0.1
Critical moisture content, % by wt: 1.5
Temperature, °C: 19

Dryer

"Wetted" area, m^2: 1.52
Warmed area, m^2: 2.1

Process

Warm water temperature in, °C: 43
Warm water temperature out, °C: 41
Pressure, mbar: 20
Water boiling point at 20 mbar, °C: 17.5
Vapor temperature in first drying period, °C: 20.5
Product exit temperature, °C: 40

First drying period
Mass balance product, kg

	In	Out
solid	138	138
H$_2$O	$\dfrac{12}{150}+$	$\dfrac{2.1}{140.1}+$

Evaporation $12 - 2.1 = 9.9$ kg

Time 2 hr.

According to Fischer (1963) the effective product temperature

is $\dfrac{20.5 + 17.5}{2} = 19°C$

Latent heat of evaporation at 19°C: 2,466 kJ/kg H_2O.

Now the heat-transfer coefficient can be calculated:

$$U = \dfrac{9.9 \cdot 2,466 \cdot 10^3}{(42 - 19)1.52 \cdot 7,200} = 97 \ W/m^2 \cdot K$$

Second drying period

The moisture content gradually drops from 1.5 to 0.1% by wt. The temperature simultaneously rises from 20.5 to 40°C. The time required is 2 hr.

Filling and homogenizing: 30 min
Discharging: 15 min
Spare time: 15 min
Cycle time: 5.0 hr
Capacity: 138/5.0 = 27.6 kg of product per hr

Scaling-up

It is assumed that a 1,000-l dryer can produce

$\dfrac{1,000}{300} \times 138 = 460$ kg of product per batch

First drying period

Time required $2\left(\dfrac{1,000}{300}\right)^{1/3} = 2.99$ hr

Second drying period: 2.99 hr
Filling and homogenizing say, 1 hr
Discharge: 30 min
Spare time: 30 min

Capacity: 460/7.98 = 57.6 kg of product per hr

So, the larger dryer can hardly produce twice the amount the smaller dryer can produce.

Sample calculation

100 kg of powdery copolymer of ethylene and vinyl acetate is warmed up from 20 to 43°C in 45 min. The 300-l cone dryer with screw is utilized, and 2.1 m^2 is covered with powder. Water of 60°C flows through the jacket.

$CP = 2{,}800 \text{ J/kg} \cdot \text{K}.$

$$U = \frac{100 \cdot 2{,}800}{2.1 \cdot 45 \cdot 60} \log_e \left(\frac{60 - 20}{60 - 43} \right) = 42.3 \text{ W/m}^2 \cdot \text{K}$$

10.5 Tumbler

In a vacuum tumble dryer, wet particulate matter can be dried by indirect heat transfer to the material while the charge is in a gentle rolling and folding motion that is due to gravity. This definition distinguishes the tumble dryer from systems having flights, vanes, plows, or scrapers that induce mobility. Unlike a vacuum tray dryer, the tumble dryer has the potential to produce a finely divided dry mix rather than a hard cake. Both the tumbler and the tray vacuum dryers have an advantage over atmospheric tray dryers in that drying can occur at a reasonable rate at relatively low temperatures.

Fischer (1963) indicated that overall heat-transfer coefficients of between 85 and 115 W/m^2 · K can be found for solids being dried whereas 6–12 W/m^2 · K is applicable for fairly dry solids.

On scaling up, the number of units of heat-transfer area per unit of volume is decisive, assuming that the process conditions (vacuum, feed, and jacket water temperature) are the same for each size. To illustrate this, on increasing the volume by a factor of 3, the area becomes approximately twice the small-scale area. These considerations set a limit to scaling up.

One might consider increasing the jacket water temperature; however, process safety or product quality may preclude this possibility. A separate temperature limitation would be the tendency of the heated product to coat the interior walls of the tumbler. Even a very thin incrustation can create a disastrously effective heat barrier.

The resistance to heat transfer is located almost exclusively in the step from the wall to the product. The jacket water temperature hardly decreases between the inlet and the outlet. Typical applications are the drying of polyester and nylon chips to a final moisture content of less than 0.01% by weight. If nylon is in contact with air at temperatures higher than 40°C, oxidation occurs. Thus, prior to discharge, indirect cooling is used to bring the maximum product temperature of 110°C (Nylon 6) down to less than 40°C.

The jacket temperature can range up to 300–350°C while the vacuum in the dryer can be as low as 1 mbar. The largest dryers have volumes of up to 30–40 m^3, the batch usually occupying up to 60% of the dryer volume. Drying cycles can be up to 20 hr.

Semitechnical testing is usually carried out in dryers having volumes of up to several hundred liters. Sometimes the tumbling action can cause

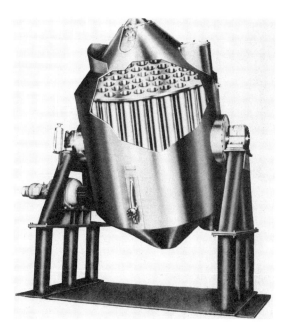

Figure 10.4 A cutaway showing 6-in diameter tubes in a Patterson-Kelley Tubular Vacuum Dryer (300 ft^3). (Courtesy of Patterson-Kelley Co., East Stroudsburg, Pennsylvania.)

balling; in such instances, the tumbler is not really suitable. Figure 10.4 illustrates an advanced tumbler. The tubes provide the heat-exchange area and the maximum size available is 8.5 m^3.

10.6   Vacuum tray dryer

A vacuum tray dryer is a contact dryer shaped like a chamber in which are placed trays containing the solid to be dried. Much of what was said concerning the atmospheric tray dryer is equally applicable here.

The heat supply is by means of warm water or other suitable heat-exchange medium, which is circulated through the metal parts on which the trays rest. The evolved vapors are absorbed by the vacuum unit. Understandably, this type of dryer has a low capacity. Specific evaporation figures vary between 0.1 and 0.2 kg/hr · m^2 of tray area. Relatively high values are found for heating plate temperatures of 150°C.

Vacuum tray dryers are used fairly extensively for temperature sensitive

Figure 10.5 A vacuum tray dryer. (Courtesy of APV-Mitchell Dryers Ltd, Denton Holme, Carlisle, England.)

or easily oxidized materials. Industrial sizes of up to several m^3 are available (see Fig. 10.5).

10.7 Agitated atmospheric dryer

In an agitated atmospheric dryer a batch of wet particulate material can be dried at atmospheric pressure by heating the agitated material indirectly so as to cause the evaporation of the moisture. When dealing with solutions, if the moisture is water and if the drying is to occur at a reasonable rate, the product temperature must be 100°C or higher.

Atmospheric dryers are still found in small production systems because of their simplicity. Much of what was stated for the agitated vacuum dryer, including typical overall heat-transfer coefficients, test work, design equations, and scaling-up practices, is equally applicable here.

Generally, one may distinguish between horizontal and vertical agitated dryers. The simplest vertical dryer is the pan-type dryer with top-driven paddle in which material cannot get trapped in seals or bearings. Vertical dryers can vary in size with a diameter of up to several meters and the diameter/depth ratio varying between 3 and 6. Low-speed agitation is practiced (several rpm).

The horizontal agitated dryers usually consist of a jacketed horizontal cylinder of length/diameter ratio 2.5 to 5. The agitators are often radial paddles, but scroll blades are also found. The largest dryers have a capacity of several tenfolds of cubic meters (see Fig. 10.6).

Figure 10.6   A mechanical pan dryer. (Courtesy of APV-Mitchell Dryers Ltd, Denton Holme, Carlisle, England)

Notation

A	Heat transfer area	m^2
CP	Solid specific heat	kJ/kg $\cdot$ K
C_p	Drying gas specific heat	kJ/kg $\cdot$ K
G_{air}	Dry air mass	kg
G_{gas}	Drying gas mass flow	kg/sec
G_{H_2O}	Evaporated water mass	kg
G_s	Solid mass	kg
TA_{IN}	Drying-air temperature	°C
T_i	Heating medium inlet temperature	°C
$T_p(t)$	Product temperature at time t	°C
t	Time	sec
t_1	Duration of first drying period	sec
t_2	Duration of second drying period	sec
t_3	Duration of cooling period	sec
U	Heat-transfer coefficient	W/m^2 $\cdot$ K

y	Air water-content	kg/kg of dry air
y_o	Drying air water-content	kg/kg of dry air
y_i	Spent drying air water-content	kg/kg of dry air
Δy	Drying air water-content increase	kg/kg of dry air

References

Entrop, W. and R. Jensen (1975). Vacuum or pressure mixing, *Process Eng.* 56, 43.

Fischer, J. J. (1963). Drying in vacuum tumblers, *Ind. Eng. Chem.,* 55, 19.

Lücke, R. (1976). Local heat transfer coefficients in a dryer with plough-type agitating elements, *Verfahrenstechnik,* 10, 774 (in German).

Stein, W. A. (1976). Drying batchwise with different vacuum contact dryers, *Verfahrenstechnik,* 10, 769 (in German).

Stein, W. A. (1977). Drying batchwise with different vacuum contact dryers, *Verfahrenstechnik,* 11, 108 (in German).

11

SPECIAL DRYING TECHNIQUES

11.1   Introduction

The following techniques are discussed in this chapter: (a) dielectric drying, (b) infrared (IR) drying, (c) freeze-drying, (d) centrifugal fluid-bed drying, and (e) slurry/paste drying. The first two methods involve drying by radiation, the third is a special type of contact drying, whereas the last two involve convective drying.

11.2   Dielectric drying

Introduction

Dielectric drying results when heat is generated by the application of electromagnetic fields, the energy source being an electric network. Because of the latter aspect, dielectric drying is often superficially considered inefficient because a conventional power station's efficiency is approximately 40%. Once network and transformer losses are also included, this figure decreases to about 33%. In many instances, however, a price/performance evaluation can show the advantages of dielectric drying.

The heat generation can be compared with that in a home microwave oven; the heat is used to prepare food but not to evaporate moisture.

A spectacular example of the benefits of microwave heating is the tempering of frozen food products to just below the freezing point. For example, it takes only two minutes for microwave heating to temper a frozen 27-kg block of meat to just below 0°C compared with up to 48 hr required by conventional thawing.

Dielectric drying offers advantages when the material being dried is not particulate (e.g., textiles) or is particulate with a large particle size (centimeters or decimeters), i.e., when the surface to volume ratio is small. Heat input by convection, conduction, or radiation is rather inefficient in these cases. Dielectric heating has the unique ability to generate heat within the product.

Dielectric drying can show the following advantages when compared with convective or contact drying: (a) lower fixed costs, (b) lower variable costs, (c) better product quality, (d) better process control, (e) clean energy source, and (f) less floor space.

An example of the use of dielectric drying is found in pasta manufacture. Conventionally, employing warm air involves a drying period of between 5 and 8 hr; the long period is required to avoid hardening and checks. Microwave drying takes only 1 to 2 hr, with initial drying down to a moisture content of 20% being carried out with warm air followed by additional microwave drying in which warm air is utilized to sweep away the evaporated moisture. The microwave drying system uses about 25% less energy than that employed in conventional dryers (Mans, 1984).

Microwave drying uses electromagnetic fields of frequency 100–10,000 MHz and radio frequencies of 1–100 MHz. Radio-frequency (RF) drying is used more often than microwave because:

1. RF has a greater penetration depth and hence is more suitable for large objects.
2. RF is less expensive an investment than microwave.
3. RF installations have much more power (the largest microwave drying system is 60 kW whereas RF dryers of 2–3 MW are common).

In order for dielectric heating to be utilized it must, of course, be possible to introduce the heat into the material being dried. This means that the material must have an electrical conductivity intermediate between conductors and insulators. The coupling of the electromagnetic field into the process stream can proceed by two mechanisms: (a) an ionic one (i.e., resistance heating) and (b) a dipole orientation. The dipole mechanism is found more frequently than the ionic mechanism. A molecule possesses a dipole when the center of the positive charge does not coincide with the center of the negative charge. A well-known example of this is the water molecule. Molecules having a dipole will be orientated by the electromagnetic field and this movement will, through friction with adjacent molecules, generate heat. Insulators (e.g., glass and Teflon) are transparent to radio waves and radar, conductors (e.g., stainless steel) reflect them, and, hence, can be used to guide the waves. Dielectric materials are acted upon

Figure 11.1 A 4,000 lb/hr short-cut pasta microwave hot-air drying system. (Courtesy of Microdry, San Ramon, California.)

(are susceptible). Figure 11.1 shows a large microwave dryer and Figure 11.2 illustrates a radio-frequency dryer.

Theory

The electromagnetic fields used are radio-frequency and microwave. Equation (11.1) is valid for both fields:

$$v = f \times \lambda \tag{11.1}$$

Figure 11.2   A cracker production unit using radio-frequency postbaking (50-kW unit). (Courtesy of Strayfield International Limited, Reading, England.)

where

   v = propagation velocity, m/sec
   f = frequency, sec^{-1}
   λ = wavelength, m

The propagation velocity in a vacuum is 3×10^8 m/sec. This is the speed at which the light of the sun travels to the earth through space. On entering a medium (e.g., a glass), the frequency remains constant. However, both v and λ vary.

   The formula expressing the heat generation that is due to a dipolar mechanism is:

$$P = 2\pi \cdot f \cdot E^2 \cdot \epsilon_0 \cdot \epsilon_r \cdot \tan \delta \tag{11.2}$$

where

P = power generation, W/m^3

E = electric field strength, V/m

ϵ_0 = dielectric permittivity of free space, i.e., $8.85 \cdot 10^{-12}$ F/m

ϵ_r = the dielectric constant of the material, i.e., the ratio of the capacity of a capacitor at a specified frequency, with the material as dielectric to that of the same capacitor in vacuo

tan δ = the loss tangent, i.e., the tangent of the phase angle between the field in the material and the applied field

The product $\epsilon_r \cdot$ tan δ is named the *loss factor*. A rule-of-thumb is that materials exhibiting, at a given set of temperature and frequency, a loss factor greater than 0.02 can be considered receptive to RF heating. Microwave heating can be effective at even smaller values of the loss factor (see Table 11.1).

The table illustrates: (a) the variation of the loss factor with the frequency, (b) the low values for plastics, and (c) the variation of the loss factor with the aggregation state (ice versus water). Equation (11.2) brings out the dependency of P on f. It will be clear that, in general, higher power densities can be attained with microwaves than with RF. In general, the electric field strength on applying RF is the potential difference between the two electrodes (plates) divided by their distance, provided the article just fits in between the electrodes. It is preferred to fill up completely the space between the electrodes with the product to be dried. If, e.g., air is also present, the field lines pass through this medium preferentially and, to a great extent, avoid the product. Often, RF heating is operated with high field strengths to compensate for the low f values. A practical limit is 100,000 V/m for most materials. Clean dry air has an electrical breakdown strength of 3,000,000 V/m.

Table 11.1 The Dielectric Loss Factors of Some Common Materials

Material	Temperature, °C	$\epsilon_r \cdot$ tanδ, −	
		At 10 MHz	At 3,000 MHz
Suet	25	4.2	0.18
Board in machine direction	ambient	0.8	0.4
Plexiglass	27	0.027	0.015
Ice, pure	−12	0.07	0.003
Water	25	0.36	12.0
Aqueous NaCl solution, 0.1 molal	25	100.0	18.0

Source: The majority of the data can be found in Von Hippel (1954).

The penetration depth of the electromagnetic waves is proportional to the wavelength. Furthermore, it depends on the material's dielectric properties. A receptive material absorbs the energy readily, and sometimes a piece gets a relatively warm exterior and a cool interior. RF waves have wavelengths of several meters, whereas microwaves have wavelengths of several centimeters. In general, RF heating has a greater penetration depth than microwaves and is more suitable for large objects. Microwaves cannot penetrate very deep but can generate much heat per unit volume.

The electromagnetic waves used for dielectric heating are the same waves that are used for radio and radar communication systems. For dielectric heating, the industrial, scientific, and medical (ISM) bands can be used. These bands were established by international agreement. Preferably, they are used for drying; however, other wavelengths can be used when shielding is adequate. As a result, 27.12 MHz is accepted worldwide for RF heating and 2,450 MHz is generally adapted for domestic microwave ovens.

Equipment

Both types of dielectric heating employ a generator to convert the main frequency of 50 Hz to the operating frequency. RF generators use an industrial triode and microwave generators use a magnetron.

Radio frequency (RF)

Dielectric tubes have a lifetime of 5,000–10,000 hr. However, the manufacturer's guarantee is usually much less. The applicator is that part of the installation in which the product is heated, which can take one of three different forms: (a) two flat metal plates, (b) a stray field electrode system, and (c) a staggered through-field electrode system.

Figure 11.3, which exhibits the first option, can be used for large objects. Figure 11.4 outlines the second option. Here, a horizontal non-uniform field passes through (usually) thin webs (up to 10 mm). Electrodes that are arranged both above and below the material can be seen in Figure 11.5. Relatively thick sheet material can be processed (third option). RF drying usually exhibits a self-limiting or leveling effect: a wet load causes an increase in electrode voltage, but a dry load will cause a power reduction.

Microwaves

After generation by means of a magnetron or klystron (lifetime 5,000–10,000 hr), the microwave energy is transported to the applicator. Usually, waveguides are employed (coaxial cable is also possible for low power). A waveguide is a hollow rectangular conduit made of metal; the dimensions match the nature of the transported microwaves.

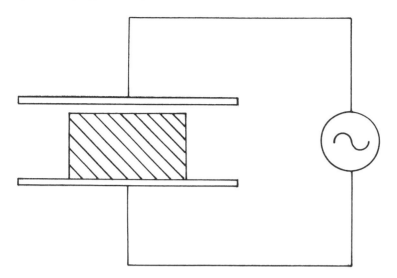

Figure 11.3 The applicator built as two flat metal plates.

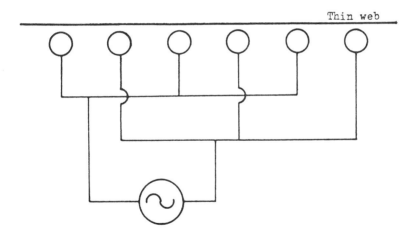

Figure 11.4 A thin web being treated with RF waves.

Waveguides can also be used as applicators, as in the passing of filamentary materials through the field in a waveguide. Traveling wave applicators are also used (see Fig. 11.6). A thin sheet of material is passed through the slots. Figure 11.7 exhibits a cavity applicator, a type used widely, such as in home microwave ovens. Basically, this applicator consists of a metal box

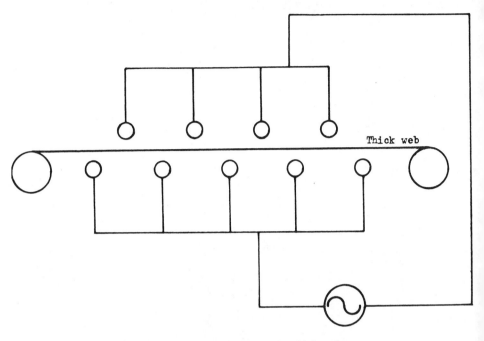

Figure 11.5 A through-field electrode system for thick webs.

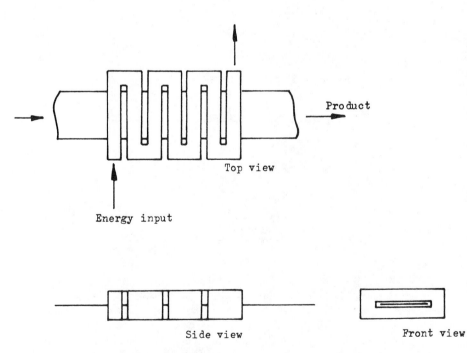

Figure 11.6 A traveling wave applicator.

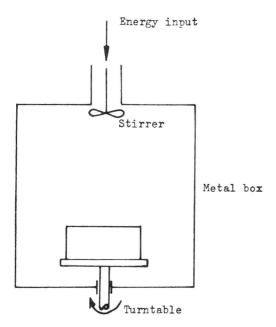

Figure 11.7 A cavity applicator.

with continuous and batchwise operations possible. It is important that the microwave field in the load be uniform. Steps to ensure this are:

1. Moving or rotating the load by means of conveyors or turntables.
2. Installing mode stirrers.
3. Using multiple ports.
4. Using multiple sources with slightly different frequencies.
5. Selecting a set of cavity dimensions.

A decrease of the evaporation load usually does not automatically lead to a decrease of the tubes' power output. In this aspect, microwave drying does not resemble RF drying. Steps must be taken to avoid overheating the tubes in case the processed material changes its susceptibility. For a continuous dryer, the leakage is controlled by reject filter traps using quarter wave choke elements or dissipative traps through which an absorbing liquid flows. Safetywise, avoiding leakages is more important for microwave drying than for RF drying. The background is the difference in wavelength.

Variables

1. *Moisture content:* Water's dielectric constant (DC) is 78; base materials have a DC of the order of 2. As the moisture content decreases, less Rf. or microwave energy couples into the processed material. This

effect will cause a moisture leveling by self-limitation. Bound water is less mobile than free water and hence the receptivity decreases (see Table 11.1).
2. *Temperature:* The power absorbed may increase or decrease when the temperature goes up.
3. *Frequency:* The frequency has an impact on the power coupled into the material to be dried.
4. *Bulk density:* Air inclusions reduce the susceptibility.
5. *Heat conductivity:* The heat conductivity of the material can be an important parameter. The heat generated must be dissipated and sometimes a practice is adopted of pulsing the power on and off to allow for an evening out of the temperature.
6. *Heat capacity:* This physical property influences the heating rate at a given power input.

Comparison of RF and of microwaves

1. RF is suitable for large objects because of its great penetration depth.
2. RF requires regular shapes, irregular shapes can sometimes be processed successfully in a microwave multimode cavity.
3. When the specific power input (W/m^3) must be high and the loss factor is smaller than 0.005, microwaves are preferable (because of limited voltage, arcing, burning).
4. Economics favor RF in the case of high power.
5. RF is preferable when the load properties fluctuate.
6. RF may be the only choice for systems containing presses, metal conveyors, etc.
7. RF can dry out paper down to 5% by wt in an even manner. Microwaves go down to 0% by wt.

Energy consumption

The evaporation of 1 kg of water at 0°C requires 2,504 kJ, which corresponds to 0.696 kWhr. Both RF and microwaves use approximately 1 kWhr for the evaporation of 1 kg of water. The consumption of electricity from the network for both types of dielectric drying is about 2 kWhr/kg of evaporated water. Heating-up is additional. Approximately 20% is added to cater for surface losses; and 2–2.5 kWhr are used per kg of evaporated water in the Belgian textile industry (RF). The drying of latex utilizes 2.3 kWhr/kg (Van Loock, 1987.)

Investment aspects and sizes

Radio frequency dryers are built up to 2,000 to 3,000 kW per drying system. Microwave dryers are built up to 60 kW. The investment figures are: (a) $1,300 to $2,700 per kW for RF and (b) $2,500 to $3,500 per kW for microwaves. Large dryers are cheaper per kW than small dryers (Schiffmann, 1987).

Microwaves

The power module for most large systems is a 60-kW magnetron at 915 MHz. Some systems have been built using multiples of small magnetrons at 2,450 mHz. The largest power magnetron available at 2,450 MHz is 6 kW.

Tests regarding continuous microwave drying are carried out in a tunnel of these dimensions: (a) width 2 ft 3 in, (b) length 11 ft, and (c) height 4 ft 10 in. The nominal processing power into the tunnel can vary from 0 to 5 kW. Frequency: 2,450 MHz. Batch experiments can be carried out in a cavity 54 in (length) × 34 in (width) × 34 in (height) (Microdry, 1988).

RF

Multiple modules are common and are well received. For example, two 50-kW units can be combined to a 100-kW dryer. Scaling up experiments can be carried out in a 25-kW unit.

Examples

Radio frequency

RF drying is often carried out by utilizing a staggered through-field electrode system. Basically, this configuration is suitable for sheet materials and thick webs. The distance between the electrodes is kept to a minimum in order to achieve heating without arcing. Postbaking of crackers is an example (see Fig. 11.2).

Water-based adhesives (e.g., PVAs, dextrines, and acrylics) are dried by means of RF techniques in the paper- and board-using industries. Both stray field electrode systems (for thin substrates up to 6 mm thick) and staggered through-field electrode systems (for thicker materials and irregular shapes) are used. The conveyor belts of the continuous dryers are made of polyester or Teflon. These construction materials do not absorb RF energy at all. The belts are often perforated to allow the passage of warm air that entrains the evaporated water. Tunnel dryers are used in the textile industry, with hanks and tops being dried by means of RF.

Microwave

The batchwise microwave drying of pharmaceutical powders under low-pressure conditions has been described (Fielder, 1983). The incentive was the wish to avoid granule attrition and hence to have a dust-free drying system that can be cleaned easily. There is a cylindrical drying chamber with the microwave generators mounted on top. A "mode stirrer" creates an even field in the drying chamber.

Continuous microwave pasta drying has been described (Hubble, 1982). A typical macaroni dryer is exhibited in Figure 11.1. It produces 4,000 lb of short-cut pasta per hr. The drying system consists of three parts:

1. A conventional hot air predryer. The dwelling time is 35 min here and the moisture content is reduced from 30% to 18% by weight.
2. The microwave-warm air stage. A further reduction to 13–13.5% occurs in approximately 12 min. The microwave system operates at approximately 30 kW. The air temperature is 82°C.
3. An equalizing stage to arrest the drying, to allow temperature gradients to equalize and to cool the product.

This dryer type is presently being adopted in the United States.

11.3   Infrared drying

Introduction

Infrared drying is accomplished by electromagnetic radiation. The wavelengths used are in the range of 1–6 microns. The drying technique is very suitable for the drying of surface coatings (paints), webs (e.g., paper and textile), and objects (e.g., ceramic tiles). Unlike dielectric drying, the penetration depth of infrared radiation is not substantial (e.g., several mm). It can offer advantages if compared to convective and conductive drying. The advantages may concern both the manufacturing cost price and the way the drying operation affects the quality of the product.

Generally, the objectives of infrared radiation are drying, hardening, or accelerating a chemical reaction. The discussion will be restricted to drying. Generally, the wavelength of IR radiation is used for classification: (a) short 0.76–2.0 micron, (b) medium 2.0–4.0 micron, and (c) long 4.0–10.0 micron.

The first commercial application in 1930 concerned the drying of cars (see Fig. 11.8). Infrared radiation can be generated in one of two ways: electrically, by passing an electric current through a resistance, or by a burning gas that heats a ceramic plate, which then emits the radiation.

The drawbacks held against dielectric drying can also be held against electric infrared drying; i.e., electric energy is expensive if compared with energy generated directly by combusting gas. However, electric drying is clean because there are no exhaust gases. The amount of air to be supplied serves to entrain only the vapor; hence, the dust problem is reduced. Gas systems are often fitted as postdryers or predryers to existing heating/

Figure 11.8 An infrared spot heater. (Courtesy of Industrial Heating Benelux, Breda, The Netherlands.)

drying systems to increase their capacity. Gas radiant heating is not capital intensive.

Theory

The infrared radiation received by a body is absorbed, reflected, or transmitted. The part of the absorbed energy is the only effective part for drying. The factors influencing the percentage of the energy useful for the drying are: (a) the properties and thickness of the coating, (b) the properties and thickness of the substrate, and (c) the geometrical configuration.

A distinction is made between black bodies, white bodies, grey bodies, and transparent bodies. A black body absorbs all radiation received (no reflection, no transmission). A black body emits a characteristic radiation. The spectral distribution of the energy is dependent on the temperature only (see Fig. 11.9). The amount of energy radiated increases when the temperature increases (areas under the curved lines). Stefan-Boltzmann's law expresses this quantitatively:

$$Q = \sigma \cdot T^4 \text{ W/m}^2 \tag{11.3}$$

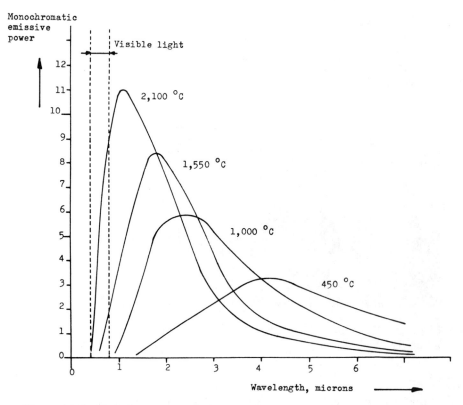

Figure 11.9   Emission spectra of a black body.

σ is Stefan-Boltzmann's constant: $5.675 \cdot 10^{-8}$ W/(m² · K⁴). The wavelength at which the maximum energy density prevails decreases when the temperature increases. This fact is known because bodies (e.g., metals) start to glow and even emit visible light when they become hot. For example, the wavelength at which the maximum energy density occurs is for the sun's light approximately 0.5 micron. Visible light ranges from 0.4 to 0.7 micron. Wien's law expresses this quantitatively:

$$\lambda_{max} \cdot T = c_w \tag{11.4}$$

c_w is Wien's constant: 2,897 μm · K.

A white body reflects all radiation received (no absorption, no transmission). Such a body cannot be heated by means of infrared radiation.

A grey body reflects part of the energy received and absorbs part of it while the fraction absorbed is independent from the wavelength. Energy is not transmitted. Many bodies preferentially absorb radiation having a certain wavelength. For example, water absorbs infrared radiation of wavelengths 2.7 and 6.0 micron preferentially. This phenomenon is quite common: the fact that a tomato is red means that light having the wavelength corresponding with red is reflected and other wavelengths are more or less absorbed. It stands to reason that it is important to find out what wavelengths are absorbed by the articles to be dried. Thus, the emission spectrum of the source and the absorption characteristics of the receiving body must be matched. Transparent bodies transmit all infrared radiation (no reflection, no absorption). Gases can often be considered transparent. However, water vapor is an exception because it absorbs infrared radiation strongly. This points to the necessity to entrain the water vapor after evaporation.

There is a difference in the drying characteristics of surface coatings on metal and surface coatings on wood. Both are usually dried by means of medium wavelength infrared radiation (2.6–4.0 microns). The surface coatings tend to absorb infrared radiation which has wavelengths of approximately 3 microns preferentially. Paints on metal do so, and the radiation, which has other wavelengths, is reflected on the metal's surface. Paints on wood behave similarly; however, the radiation passing through the surface coating is absorbed by the wood. The wood is heated thereby and, in turn, heats the paint by conduction from below. The net effect is that the paints on wood are dried homogeneously whereas surface coatings on metals may exhibit dry spots that hamper solvent release.

The energy exchanged between two black bodies can be calculated by means of Equation (11.3).

Example

Temperature first body: 800°C
Temperature second body: 75°C
The emitting and receiving boundaries are infinitely large parallel planes
The gas between the two black bodies is transparent

$$Q = \sigma (T_1^4 - T_2^4) = 5.675 \times 10^{-8}[(800 + 273)^4 - (75 + 273)^4] = 74{,}393 \text{ W/m}^2$$

It will be difficult to get this value in actual practice. Ideal black bodies do not exist, and there will be so-called convective losses (gas between the two bodies that is heated and escapes). Stefan-Boltzmann's law is modified for grey bodies into:

$$Q = \epsilon \cdot \sigma \cdot T^4 \text{ W/m}^2 \qquad (11.5)$$

ϵ is the emission coefficient. Most dull and rough surfaces have ϵ values of about 0.6–0.8, but bright metal surfaces have low emissivities in the range

0.01–0.1. Kirchhoff's law states that at each wavelength the emission coefficient of a grey body equals the absorption coefficient ($\alpha = \epsilon$). It can be proved that the heat transfer between two parallel planes (grey bodies, no transmission) can be expressed as follows:

$$Q = \frac{\sigma(T_1^4 - T_2^4)}{\frac{1}{\epsilon_1} + \frac{1}{\epsilon_2} - 1} \text{ W/m}^2 \tag{11.6}$$

The gas between the two planes must be transparent and the radiation that is not absorbed is reflected diffusely (i.e., in all directions). The two planes must be very large (to prevent reflected radiation from escaping).

A grey body can be convex and completely surrounded by a different body. Then, the following expression applies:

$$Q = \frac{\sigma(T_1^4 - T_2^4)}{\frac{1}{\epsilon_1} + \frac{A_1}{A_2}(\frac{1}{\epsilon_2} - 1)} \text{ W/m}^2 \tag{11.7}$$

If $A_2 \gg A_1$ and if ϵ_2 is close to 1

$$Q = \epsilon_1 \cdot \sigma(T_2^4 - T_1^4) \qquad \text{W/m}^2 \tag{11.8}$$

Approximately 90% of the energy consumed by an electric infrared generator is converted into radiation. The percentage for gas combustion is about 60 for advanced burners. Convection and conduction losses come extra.

Theory today does not mean too much for the design of infrared radiation equipment. Carrying out small-scale experiments to obtain design data is common. These tests produce drying times, energy consumption data, and an insight into product quality. So-called temperature/time curves are obtained (see Fig. 11.10).

An important aspect is the time delay on starting and stopping: (a) 0.76 to 2.0 microns: sec, (b) 2.0 to 4.0 microns: approximately one min, and (c) 4.0 to 10.0 microns: approximately ¼ hr. A long time delay affects process control. Furthermore, a shutdown can mean that the radiation continues for some time. Often, it is necessary to provide facilities that isolate the source from the object (swing-out arm, barrier).

Equipment

Electrical equipment

The infrared radiation is obtained by passing an electric current through a resistance. The resistance heats up and thereby attains a high temperature. Lamps and tubes are known examples. Figure 11.11 shows a cross-section of a typical tube emitter. Tungsten and stainless steel are being used as construction materials for resistances. The stainless steel emitters are pre-

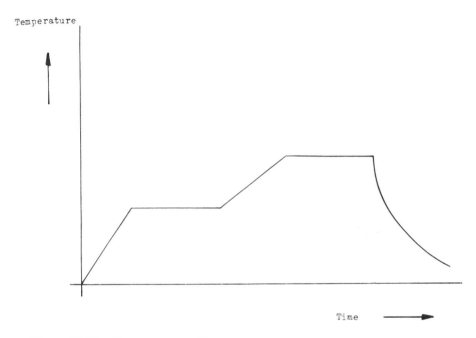

Figure 11.10 The temperature/time curve.

ferred for drying purposes and are in use for textile and for paper. Tungsten is oxidized by air at approximately 2,000°C (see Table 11.2). Electrical oven systems usually have no size limit and are tailored to a client's requirements.

Gas-fired equipment

A perforated ceramic plate is commonly employed, with the gas/air mixture flowing through the small channels and igniting (see Fig. 11.12). Gradually, as the ceramic plate (thickness, e.g., 12 mm) heats up, the flames with an approximate length of 2.5 mm travel back into the small channels. This causes the surface to glow (850–950°C) and to emit infrared radiation with wavelengths in the range 2.7 to 2.3 microns. The plate's design (porosity) guarantees a temperature of the other side of the plate of 200–250°C. This temperature is too low to ignite the gas/air mixture, thus, the combustion process is safeguarded. Approximately 60% of the energy supplied to the burner is converted into radiation energy (basis: lower calorific value). The remaining 40% comprise the losses. The exhaust gas represents the most important loss, the other losses are caused by convection and conduction. A characteristic figure is 50 kW/m^2 radiation energy. Note that on using gas-fired equipment, the product comes into contact with exhaust gases.

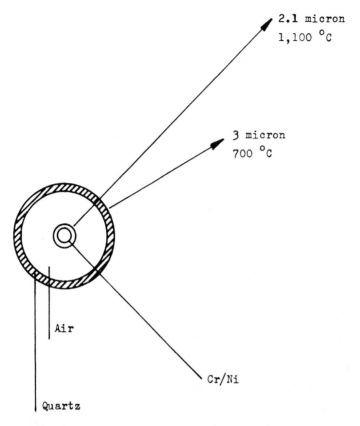

2.1 micron
1,100 °C

3 micron
700 °C

Air

Cr/Ni

Quartz

Figure 11.11 A cross-section of a typical tube emitter.

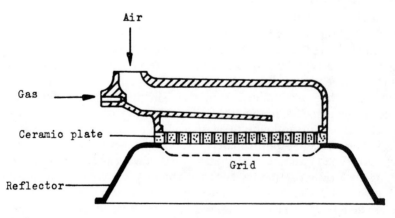

Air

Gas

Ceramic plate

Grid

Reflector

Figure 11.12 A gas-fired infrared heater.

250

Table 11.2   The Main Characteristics of Electrical Equipment

Indication	Lamp	Lamp	Tube	Tube	Tube
Material resistance	W	W	Cr/Ni	Cr/Ni	Cr/Ni
Temperature resistance, °C	1,900	2,100	1,100		
Wavelength resistance, μm	1.3	1.2	2.1		
Material wall	Quartz	Quartz	Quartz	a	a
Temperature wall, °C	170	400	700	400/800	650
Wavelength wall, μm			3	2.7–4.3	3.1
Intermediate material	Vacuum	Gas	Air	b	b

aStainless steel.
bCeramic powder.

Process control is not perfect, a turn-down to about 60% of the nominal capacity is the limit. Process control is therefore often executed via the on/off mode for groups of burners. The catalytic burner is employed for leather drying. The gas reacts with oxygen without the emission of visible light, and the surface temperature is approximately 400°C. Propane and butane combustion is more complete than methane combustion. Gas oven systems usually have no size limit and are tailored to a client's requirements. Gas heaters fitted as pre-dryer or postdryer to an existing drying system are often standardized. Fig. 11.13 shows a typical application of a gas-fired IR unit.

Testing

Successful tests provide the basis for an infrared drying system's design. Static or moving drying trials are carried out at the equipment manufacturers' pilot plants. It is also customary to carry out site trials using a portable machine. A theoretical analysis can provide a feel for the direction to look for, however, tests are a prerequisite to a good decision.

Process safety

Historically there has been some concern in industry regarding the potential fire hazard of radiant heaters. It was stated that infrared radiation equipment emitting in the wavelength range of 2–4 microns remains hot for approximately one min in the event of sudden stoppage. The material being processed may catch fire. Conventional heat shields, mechanical heater rotation, and retraction techniques have been used to decrease the risk, but these have not been fail-safe.

Stordy Combustion Engineering Ltd. (England) manufactures gas heaters that can be turned off and touched within three to four sec. The initial

Figure 11.13 Gas-fired infrared unit on a vinyl wallpaper embossing machine. (Courtesy of Stordy Combustion Engineering Ltd., Wombourne, England.)

temperature was 870°C. Rapid modulation within the operating range to match the thermal load is possible.

Another aspect is the possibility that an evaporated solvent may catch fire. For a typical range of solvents used in paint stoving applications, 60 m^3 of fresh air at 16°C is required per liter of solvent evaporated (BNCE, 1985). The power supplies to the heaters should be cut off in the event of a reduction or loss of the air supply.

11.4 Freeze drying

Introduction

Freeze-drying is a drying method with which the liquid in the solid/liquid mixture is frozen by cooling and is subsequently caused to sublime. The liquid to be evaporated is usually water; however, solvent evaporation can also be met. Freeze-drying is also termed lyophilization (from the Greek "made solvent loving"). Two broad categories of products are freeze-dried: pharmaceutical products and food. The justification for employing lyophilization is that, although different from drying in the conventional sense, the nature of the product is hardly altered. For example, denaturation (the decomposition of proteins) and loss of flavor occur in hot air drying of food but are prevented from occurring in freeze-drying. Both spray-dried coffee and freeze-dried coffee are available on the market and the tastes can be compared. Normally, freeze-dried products can be rehydrated almost perfectly.

Lyophilization is usually carried out under vacuum because the process proceeds faster at low pressures than at atmospheric pressure. Freeze-drying equipment suppliers are also vacuum equipment manufacturers. A trivial example of freeze-drying at atmospheric pressure is wet laundry drying outside on a cold winter day.

Freeze-drying of pharmaceutical products usually occurs batchwise, because of the necessity to maintain good manufacturing practices (GMPs). Often, food is dried continuously in an intermittent way. In the food field, the freeze-drying of coffee and soup mixes has practical significance. Two aspects prevent the method from extensive use: the drying method is very expensive and an efficient frozen-food distribution system exists in the Western world.

Lyophilization of pharmaceutical products is more important than the freeze-drying of food. Here, the method often cannot be replaced by a different type of drying and, furthermore, pharmaceutical products are often quite expensive. In this field, one deals with unstable delicate biological materials (e.g., human and animal vaccines, blood proteins, complex antibiotics, and vitamins). Industrial freeze-drying was introduced for the preservation of blood plasma during World War II.

The successive steps of systematic freeze-drying are: (a) preparation and pretreatment, (b) prefreezing to solidify the water, (c) primary drying at which the ice is sublimed under vacuum, (d) secondary drying to desorb residual moisture under high vacuum, and (e) packing after the vacuum has been broken with a dry inert gas. The sterilization of vitamins is an example of this type of preparation.

Freeze-drying pharmaceuticals

Preparation and pretreatment

Pharmaceutical freeze-drying may comprise sterilization, filtration, or the addition of excipients. Excipients may protect the material during freezing or drying or may aid in the reconstitution. Alternatively, they serve other purposes.

Pre-freezing The aqueous solutions are contained in vials, ampoules, or bottles. The methods to cool utilize: (a) cold shelves (at, e.g., −50°C), (b) alcohol baths (also at, e.g., −50°C), or (c) liquid N_2 baths (at approximately −195°C). Figure 11.14 depicts an industrial freeze-dryer schemati-

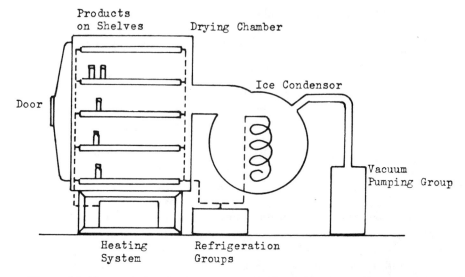

Products
on Shelves   Drying Chamber

                                        Ice Condensor

Door

                                                                    Vacuum
                                                                    Pumping Group

        Heating   Refrigeration
        System   Groups

Figure 11.14   An industrial freeze-dryer for pharmaceuticals. (Courtesy of Edwards High Vacuum International, Crawley, England.)

cally. Prefreezing, primary drying, secondary drying, and packing can all be carried out in the same drying chamber. The prefreezing then occurs on cold shelves. Normal cooling rates are in the range 1–3 K/min.

Today, freeze-drying of pharmaceuticals is often controlled by means of microprocessors. The cabinet pressure at the prefreezing stage is atmospheric, and the constant low shelf temperature is controlled. Two important aspects regarding the prefreezing are the minimum temperature that must be attained and the freezing rate.

The minimum temperature is assessed by means of a phase diagram that indicates the temperature at which on cooling ultimately all liquid has turned into solid. This so-called eutectic point can be quite low (e.g., −56°C for a solution containing calcium chloride). A low freezing rate implies the formation of relatively large ice crystals and vice versa. Relatively large pores are formed on subliming large ice crystals. This facilitates the vapor transport through the material processed.

The freezing rate is a function of:

1. The vial or ampoule dimensions.
2. The cooling medium's temperature.
3. The cooling medium's nature and the way it is applied.
4. The solution characteristics (water content, viscosity, and so on).

Thijssen and Rulkens (1969) measured mean pore diameters on freeze-dried dextrin solutions. Values in the range of 1–4 microns were found. The freezing rate varied between 25 and 3 K/min.

The phase diagram also exhibits the existence of polymorphology and metastable phases. It is important to know about polymorphology because the reconstitution attainable depends on the morphology obtained on freeze-drying. A very convenient tool for the investigation of prefreezing is the measurement of the electrical resistance as a function of the temperature. The measuring cell is exhibited in Figure 11.15. A typical record is shown in Figure 11.16. The resistance increases on cooling points to the formation of ice crystals (Willemer, 1977).

Primary drying The primary drying step occurs in the drying chamber. After the prefreezing step, time has elapsed and the microprocessor activates the vacuum pumps and keeps the absolute pressure at 0.5 mbar. The shelf temperature has in 2 hr linearly increased to 30°C. Subsequently, this temperature may be kept constant during the primary and the secondary drying steps.

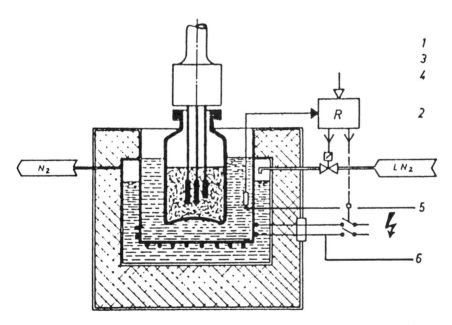

Figure 11.15 Measuring the resistance as a function of the temperature. (Courtesy of Leybold AG, Cologne, Federal Republic of Germany.)

Resistance, kΩ

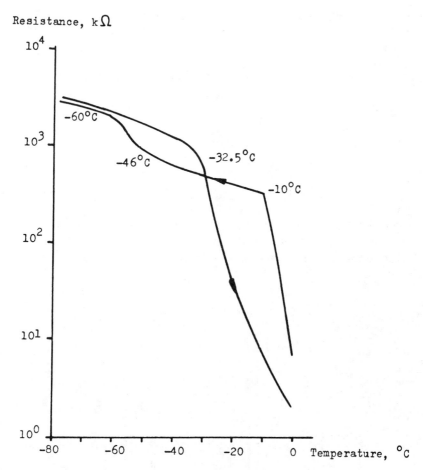

Figure 11.16   Electrical resistance of a sample on cooling and reheating. (Courtesy of Leybold AG, Cologne, Federal Republic of Germany.)

The ice saturated-vapor pressure at 0°C is approximately 6.1 mbar. At −20°C, a value of 1.0 mbar prevails, which means that the driving force for sublimation is always modest. The subliming ice front's temperature must be 5–7 K lower than the lowest eutectic temperature. The sublimed ice resolidifies on a cold metal surface. Factors that cause slow drying are: (a) the small driving force for sublimation, (b) the modest heat source temperature, (c) the necessity to keep the ice front temperature 5–7 K below the lowest eutectic point, and (d) the poor heat conductivity of the porous organic material. Typical total drying times range from 10–20 hr. The pressure rise method is a reliable end-point detection method. It comprises an isolation of the drying chamber and the vacuum pumps. Subsequently, the pressure rise is measured. If the pressure rise is more than 3% of the original value, the primary drying is continued. The first observation is that the pressure measured after the isolation is the subliming ice front's saturated vapor pressure and, therefore, the ice front's temperature is known (barometric temperature measurement). The second observation is that a relatively large pressure increase in the drying chamber isolation points to a relatively large vapor flow (associated with a certain pressure loss). If the material gets "dry" the vapor flow that is resorbed decreases.

Secondary drying   As soon as the microprocessor senses that the pressure increase on isolation is small, the secondary drying stage is initiated. The shelf temperature is kept at 30°C and the vacuum is rapidly lowered to 0.07 mbar. "Difficult" moisture is now sublimed. Again, the end-point detection is determined by measuring the pressure rise on isolation.

Packing   Closing the vials or bottles occurs in the drying plant after the vacuum has been broken with a dry inert gas.

Freeze-dried pharmaceuticals can be quite expensive. The process conditions for a large freeze-dryer are experimentally established by means of a pilot freeze-dryer (with a size, e.g., of 0.6–1 m^2). Figure 11.17 shows an intermediate-scale freeze-drying plant and Figure 11.18 exhibits a production unit.

Freeze drying food

Preparation and pretreatment

The materials to be dried should be in a physical form having a large specific area and should be evenly distributed throughout the drying chamber. Often, the foods are cut or ground; others are blanched (a treatment that stops biological activity) or precooked. Sulphur dioxide may be used to inhibit browning reactions. Usually, fruit juices are concentrated.

Prefreezing

Large industrial freeze-drying plants contain a freezing room that is separated from the dryer itself. Two different methods to freeze can be distinguished: on drums or belts (contact cooling) and on trays by cold air circulation (convective cooling). As for pharmaceutical products, the freezing process has a great influence on the quality of the product. Again, the two important independent variables are the freezing rate and the minimum temperature.

The refrigeration rate has a great influence on the structure, the consistency, the color, and the aroma retention. A slow rate promotes the formation of large ice crystals. Large crystals lead to large pores promoting sublimation; however, large crystals can damage the cells. A slow rate leads to

Figure 11.17 An intermediate-scale freeze-drying plant for pharmaceuticals. (Courtesy of Leybold AG, Cologne, Federal Republic of Germany.)

Figure 11.18 An industrial lyophilizer having 20 m^2 shelf area before installation. From left to right: the drying chamber-cylindrical ice condenser-refrigeration system. The front of the drying chamber will be sealed into the wall of the sterile area. (Courtesy of Edwards High Vacuum International, Crawley, England.)

relatively large solid particles between the large ice crystals. Thus, aroma retention is promoted. The freezing process leads to ice crystals. The remaining solution is concentrated and as the process proceeds, concentration gradients between the intracellular and the extracellular solutions develop. These gradients may affect the quality adversely, and to avoid this quick freezing is recommended. Usually, freezing rates are in the range of 0.5–3 cm/hr. The minimum temperature can be determined by a eutectic point. However, the last traces of the water in the solution do not have to be frozen.

Primary drying

Figure 11.19 depicts a continuous five-tunnel freeze-dryer. Batch dryers also exist. The prefrozen materials are loaded on trays and the feed subsequently enters the dryer via an entry lock. The trays on entering the dryer are connected to an overhead rail to enable transport through the dryer and contact heating. Heating can be effected by means of oil, water, or vacuum steam. Vacuum steam is particularly suitable because it gives a uniform serviceside temperature (temperature variations between 45 and 120°C are possible). The chamber pressure is in the range of 0.1–1 mbar. About 50–80 kg of fresh products can be dried per m^2 and per 24 hr on utilizing a special tray design (Leybold, 1975). The dried material leaves the dryer via an exit lock. The trays (combined to a carrier) are washed and recycled.

Normally, postdrying is carried out until the residual moisture is in the range of 2–3% by wt. Figure 11.20 shows the front view of a tunnel section and Figure 11.21 shows a lock to a tunnel with a vacuum-tight gate valve in the background.

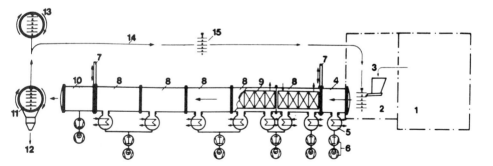

Figure 11.19 A flow diagram of a continuous five-tunnel continuous quality control freeze-drying plant. 1, Freezing room; 2, loading room; 3, tray-loading station; 4, entry lock; 5, condensor; 6, pumping system; 7, gate valve; 8, freeze-drying tunnel; 9, heating-plate system with CQC suspended carriers; 10, exit lock; 11, unloading station; 12, product discharge; 13, washing equipment; 14, rail system for carrier transport; 15, CQC suspended carriers. (Courtesy of Messrs. Leybold AG, Cologne, Federal Republic of Germany.)

Figure 11.20   A front view of a continuous quality control tunnel section. (Courtesy of Leybold AG, Cologne, Federal Republic of Germany.)

Typically, the product is in direct contact with the heating surface. The heat required for the sublimation passes through the material to be processed and then to the ice front. Part of it flows through the material already dried, and part through the gas in the pores. The sublimed water vapor flows through the pores from the ice front to the drying chamber. If it is assumed that the heat supply is abundant and that it is difficult for the water vapor to escape, such a situation would lead to a melting of the ice front. Here, the remedy is to control the heat supply carefully to avoid melting, which is a situation controlled by mass transfer. The reverse situation is that it is difficult for the heat to penetrate to the receding ice front and that the sublimed water can leave freely. Thus, we have a situation controlled by heat transfer.

The heat conductivity of the water vapor in the pores and in the dried material is low. In actual fact, the prevailing figures can be compared to the figures that are valid for insulators. These low conductivities and the possibility for water vapor to escape freely at low pressure cause the industrial freeze-drying process to be controlled by heat transfer.

Figure 11.21   A tunnel freeze-dryer lock with a gate valve in the background. (Courtesy of Leybold AG, Cologne, Federal Republic of Germany.)

Secondary drying

Bound water is removed at this stage. Neither the temperature of the heating system nor the pressure in the drying chamber is different from the temperature and pressure at the primary drying. The secondary drying stage is entered after a certain dwelling time.

Packing

The freeze-dried products are packed in a manner to prevent moisture pick-up.

The method for freeze-drying foods described so far is called the *static-drying method*. This indication refers to the fact that the freeze-dried product remains stationary on trays (compare a box oven).

A dynamic drying method can also be distinguished. The product passes from top to bottom through a piece of equipment in which dynamic drying occurs. The heating surface is also a vibratory conveyor. The feed must be free-flowing and granular. High specific evaporation rates are achieved (e.g., 5 kg/m^2 · hr). It is possible to obtain a desired temperature profile by means of different heating loops. Continuous product flow is obtained by means of entrance and exit locks. Drying times between 6 and 9 min can be registered for granules having a size between 0.5 and 2 mm and a water content of 60%. An example is the continuous freeze-drying of coffee that is subsequently packed in small packages to be used, e.g., in hotel rooms. The capacities of the continuous tunnel dryers range, depending on the type of feed, from 5 tons of intake per 24 hr to 20 tons. Here too, the optimum freezing and drying conditions are assessed empirically by means of pilot plant drying (size, e.g., 0.6–1 m^2).

11.5   Centrifugal fluid-bed drying

Introduction

It is convenient to compare centrifugal fluid-bed drying to conventional fluid-bed drying and conveyor drying. The latter two techniques are being carried out in the earth's field of gravity. Thus, the air velocities employed must be moderate to prevent the particles from being removed from the drying chamber. The radial acceleration in a centrifugal fluid-bed dryer is one to two orders of magnitude larger than the acceleration that is due to gravity. Consequently, the drying air velocities can be much greater than those in use for conventional fluid-bed dryers and for conveyor dryers. These greater air velocities lead to greater product throughputs if the drying is controlled by heat transfer (first drying period). Alternatively, the specific throughput rates (kg/m^2 · hr) can be increased considerably.

Two types of equipment will now be discussed: APV Pasilac's machine and Krauss Maffei's construction.

APV Pasilac Centrifugal Fluid Bed (C.F.B.)

The main application for this unit is the rapid initial drying of particulate material, mainly food products.

The predryer can then be followed by conventional low air-velocity type dryers to remove the moisture when the drying is controlled by mass transfer (second drying period). Figure 11.22 shows a side view of the C.F.B. cage, which is encased in the drying chamber. Hot air is introduced through the upward moving side of the perforated cylinder at such a velocity that the adjacent particles become fluidized. Air leaves through the downward moving side of the perforated cylinder passing through the fixed bed of particles. Thus, the hot air makes contact with the material twice as it passes through the drying zone.

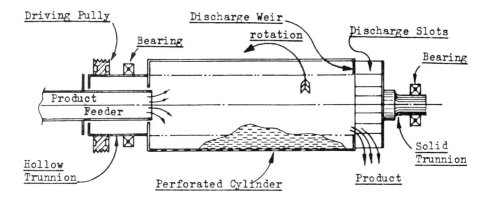

Side view of the C.F.B. cage which is encased in the drying chamber.

Figure 11.22 A centrifugal fluid-bed dryer. (Courtesy of APV Pasilac Limited, Carlisle, England.)

Typical figures
Machine data

Cylinder diameter: 0.914 m (3 ft)
Cylinder length: 6.401 m (21 ft)
Rotational speed: 100–120 rpm
Process data

Air velocities: 7.5–15 m/sec
Air temperature: 100°C

A rotational speed of 100 rpm corresponds with a radial acceleration of approximately 50 times greater than the acceleration that is due to gravity.

The centrifugal fluid-bed dryer will uniformly heat almost any shape of particle, such as vegetable dice, sliced or leaf materials, and it is not a prerequisite that the material to be dried should be capable of fluidization in a conventional fluid-bed dryer. Even odd-shaped materials, such as potato chips, have been handled successfully. The background is the tumbling action within the fluidized zone.

The dryer can be used as a continuous "puffing" device. Feeds containing 20–40% by wt of moisture are treated in the dryer. The high rate of heat input causes water to evaporate within the particles at a rate exceeding the rate at which it can pass through the surface skin. This causes the

Table 11.3 Typical Feed and Product Moisture
Contents for the C.F.B. Dryer

Material	Feed MC, % by wt	Product MC, % by wt
Bell peppers	91	83
Cabbage	89	80
Carrots	90	82
Green beans	89	80
Onions	81	68
Garden peas	82	70
Potatoes	80	67

Source: APV Pasilac Ltd, Carlisle, England.

particles to swell and produce a honeycomb structure. Puffed dried vegetables rehydrate much more readily than conventionally air-dried materials.

Krauss Maffei Centrifugal Fluid-Bed Dryer

The main application for this dryer is the rapid drying of particulate material having an average particle size in the range 50–500 microns. The moisture to be evaporated must be free moisture mainly. Figure 11.23 shows a cross-sectional view of this machine. The radial acceleration is 20 to 80 times greater than the acceleration that is due to gravity. The air velocity can be as high as 6 m/sec and the air temperature is 300°C maximum. Possible product throughputs range from 2 to 40 tons/hr. Test drying is carried out on a 0.08-m^2 pilot dryer. Typically, the dryer can be applied for rubber accelerators, detergents, plastics, and silica.

11.6 Slurry/paste dryer

Fig. 11.24 shows the principle of this dryer. The feedstock, a slurry, solution, or paste is sprayed in at the bottom of the dryer. The material adheres to the bed particles and is carried to the top of the bed during which time it is dried. The bed particles collide with the impact frame, the dryer walls, and with each other, and this causes the dried material to dislodge. In a way, the slurry/paste dryer is a normal convective dryer that in many aspects can be compared with a spray dryer. However, it occupies much less space and is thus a lower investment.

Normal hot-air inlet temperatures are in the range of 175–225°C. The inlet air velocity is 6 m/sec. This dryer has potential for special applications (such as, fine china, baby foods, and metal process sludges).

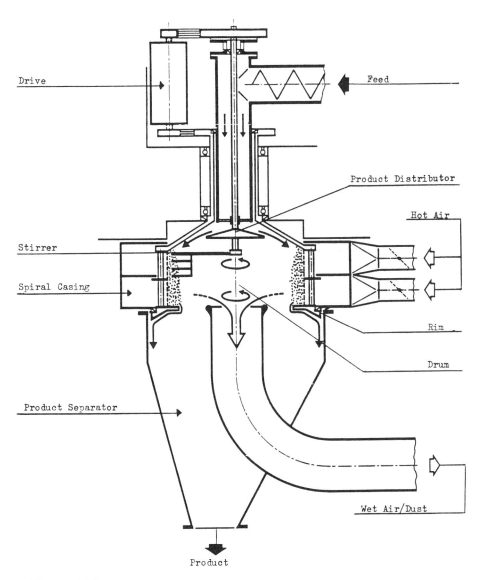

Figure 11.23 A centrifugal fluid-bed dryer. (Courtesy of Krauss Maffei, Munich, Federal Republic of Germany.)

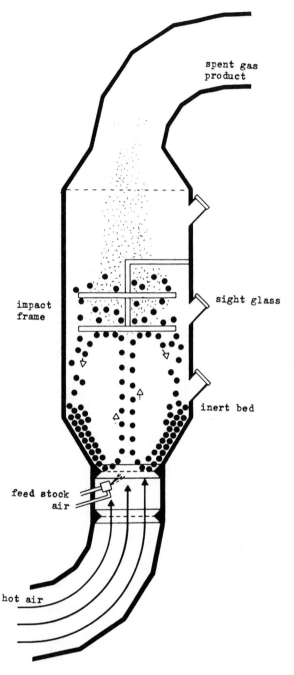

spent gas
product

impact
frame

sight glass

inert bed

feed stock
air

hot air

Figure 11.24 A slurry/paste dryer. (Courtesy of Euro-Vent, Stoke-on-Trent, England.)

Figure 11.25 An installed slurry/paste dryer. (Courtesy of Euro-Vent, Stoke-on-Trent, England.)

The inert bed material is silicone rubber, polypropylene, or Teflon in cylindrical or spherical shapes. Gas-solid separation is usually by means of a cyclone followed by a scrubber.

Dryers utilizing the same principle but operating with lower air velocities also exist. The inert bed particles are then transported by one or more screws. However, screws add to the complication of the drying process. Figure 11.25 shows an installed slurry-paste dryer.

Notation

A_1	Area of body No. 1	m^2
A_2	Area of body No. 2	m^2
c_w	Wien's constant (2,897)	$\mu m \cdot K$
E	Electric field strength	V/m
f	Electromagnetic field frequency	1/sec
P	Specific power generation	W/m^3
Q	Energy density	W/m^2

T	Temperature	K
T_1	Temperature of body No. 1	K
T_2	Temperature of body No. 2	K
v	Electromagnetic field propagation velocity	m/sec
α	Absorption coefficient	
ϵ	Emission coefficient	
ϵ_0	Dielectric permittivity of free space (8.85×10^{-12})	F/m
ϵ_1	Emission coefficient of body No. 1	
ϵ_2	Emission coefficient of body No. 2	
ϵ_r	Dielectric constant	
δ	Loss angle	
λ	Electromagnetic field wavelength	m
λ_{max}	Wavelength of maximum energy density	μm
σ	Stefan-Boltzmann's constant (5.675×10^{-8})	W/(m$^2 \cdot$ K^4)

References

BNCE (British National Committee for Electroheat) (1985). *Infra-red process heating,* BNCE, London.

Fielder, T. (1983). Microwaves—a new way to dry drugs, *Manufacturing Chemist* [England] 54, 45.

Hubble, P.E. (1982). Consider microwave drying, *Chem. Eng.,* 89, 125.

Leybold AG (1975). *Industrial Freeze Drying (27D7.100e),* Leybold AG, Cologne.

Mans, J. (1984). Computerized pasta plant, *Prepared Foods* [U.S.A.], 153, 87

Microdry (1988). Private communication.

Schiffmann, R. (1987). Dielectric drying, in *Handbook of Industrial Drying* (A.S. Mujumdar, ed.), Marcel Dekker, New York.

Thijssen, H. A. C. and W. H. Rulkens (1969). Effect of freezing rate on rate of sublimation and flavour retention in freeze-drying, Institut International du Froid, *Transactions Committee X Meeting at Lausanne [Switzerland] on June 5 and 6, 1969.*

Van Loock, W. M. A. (1987). Some applications of electromagnetic energy in Belgium, *Transactions Symposium on Dielectric Drying at Arnhem* [The Netherlands] (Organized by ETECE) (In Dutch).

Von Hippel, A. R. (ed.) (1954). *Dielectric Materials and Applications,* Technology Press of M.I.T., Cambridge, Massachusetts and John Wiley & Sons, New York.

Willemer, H. (1977). The condition of aqueous solutions during freezing demonstrated by electrical resistance measurements and low-temperature freezedrying, *Transactions Joint Meeting I.I.R.-Commissions C1 and C2 at Karlsruhe* (Federal Republic of Germany).

12
SAFEGUARDING DRYING

12.1 Introduction

During drying operations, the principal hazards are fires, explosions, and decompositions. Drying operations should be safeguarded to prevent personnel injury, to prevent plant damage, and in order to remain a reliable manufacturer.

A fire is a chemical reaction with oxygen in which light, heat, and flame are produced. Fires can occur whenever a combustible material is being dried. If a combustible solvent is being evaporated rather than water, the likelihood of a fire increases. Since many dryers are directly fired, a further possibility is the uncontrolled combustion of the heating medium (e.g., gas or oil) leading to the spreading of flames and ignition of the material being dried.

An explosion is a fast release of energy and, for drying, the two relevant types of explosions are dust and vapor explosions. Both are fires that propagate rapidly and, in a confined space, cause pressure rise. A pragmatic classification of explosions is as follows:

1. A physical explosion is the bursting of a container because of overpressure in the container. Examples include boiler explosions and vessel destruction because of high-pressure steam that is admitted to a coil or jacket designed for low- or medium-pressure steam.
2. A thermal explosion occurs when an exothermic reaction (usually in a liquid) generates heat faster than it can be dissipated, thus raising the reaction temperature to a point at which the reaction rate becomes

catastrophic and the reaction vessel is torn apart. An example of a thermal explosion occurred at Seveso (Italy) and resulted in the contamination of the environment by dioxin.

3. A nuclear explosion can be brought about by many causes but typically will result in contamination of the environment by nuclear radiation.
4. A vapor explosion occurs when an air/combustible gas mixture or an oxygen/combustible gas mixture is ignited. The combustible gas concentration must be between the lower- and upper-explosion limits.
5. A dust explosion occurs when an air/explodable dust mixture or an oxygen/explodable dust mixture is ignited. The dust concentration must be between the lower- and upper-explosion limits.
6. Explosions of liquids and solids can occur when certain materials (such as lead azide) are subjected to shock or impact.

These types of explosions (with the exception of physical explosions) can manifest themselves as deflagrations or detonations, the distinguishing feature of which is the rate at which the reaction-front propagates. In the discussion that follows, dealing with deflagrations and detonations will be restricted to vapor and dust explosions because these are of particular relevance to drying operations.

Deflagration

The word *deflagration* is derived from the Latin *deflagrare* meaning "to burn down." Burning, even of solids, requires the reacting components to be in the gaseous state. A deflagration is a fire in which a flame front travels rapidly but at subsonic speed through a gas. Typically, on confinement, a deflagration causes pressure to rise from the initial value to up to eight times the initial pressure. The explosions that can occur in drying systems fall almost exclusively within this category, and this is true for both dust and gas explosions in dryer systems.

In the case of dust explosions, the flame travels through a suspension of the dust in a gas and not a pure gas. Although this is true, prior to combustion, the solid particle must sublime.

Gas-phase detonations are known and might occur in dryers. Experience has shown that gas-phase detonations usually start out as deflagrations. However, the conversion into detonations requires path lengths exceeding drying equipment sizes. An exception to the latter is that conversion may occur in a line with a large L/D ratio. The same conversion can, in principle, be observed for dust explosions (Bartknecht, 1978).

The pressure exerted by a deflagration in a confined space can cause considerable damage. A peak pressure of 8 bar can be obtained if the initial pressure is 1 bar, whereas drying equipment has a comparatively low strength, typically yielding at 1.2 bar. However, the rate of pressure rise is also an important factor.

Detonation

In the gas phase, a detonation is defined as a deflagration having a flame velocity greater than the speed of sound. Condensed phase detonations can also occur, however; one then deals with explosives. The drying of this class of materials is not covered in this chapter. The instant and complete decomposition of lead azide is an example of detonation where the decomposition can be initiated by heat, shock, or impact.

Typically, confined gas-phase detonations cause much higher pressures than confined deflagrations; peak pressures of 20 to 40 bar (initial pressure 1 bar) are mentioned in the literature. The rates of pressure rise for confined gas-phase detonations far exceed those experienced for deflagrations. Examples of detonatable gas mixtures are hydrogen plus oxygen and hydrogen plus chlorine.

The possibility of a gas-phase detonation in a dryer can in some instances not be excluded. An example would be the evaporation of carbon disulphide on drying convectively with air; here inerting with, for example, nitrogen is strongly recommended.

The drying of materials liberating copious amounts of gas on decomposition has also been excluded from this chapter because special measures are required for these materials.

The process safety during drying can be evaluated by measuring product-safety characteristics of the material since, under normal circumstances, a chemical reaction does not occur.

Principally, two different methods for dealing with the risks of fires and explosions can be distinguished: prevention and cure. Safeguarding by means of prevention relies on process conditions excluding the possibility of undesired events occurring. Curing comprises minimizing the effect of a fire or explosion. All things being equal, prevention is to be preferred to curing; however, sometimes a balance has to be struck between cost and efficiency. See Gerritsen and Van 't Land (1985).

Finally, it must be noted that dust explosions and fires can give rise to a series of events; for example, a dust explosion may result in a fire which, in turn, may cause another dust explosion, and so on.

12.2 Fires

Many of the materials that are dried are flammable and at the elevated temperatures prevailing and in the presence of oxygen they may take fire spontaneously (self-ignition) or they may be ignited. There are three distinct types of fires for the materials being processed: (a) ignition of dust clouds, (b) ignition of dust layers or deposits, and (c) ignition of bulk materials.

Because of the high rate at which the flame front propagates during the combustion of a dust cloud, the process is known as a *dust explosion*. This topic is dealt with in Section 12.3. Burning of both dust layers and bulk dust are generally considered to be fires.

IChemE (1977) discussed two flammability screening tests for layers of powder, the first of which was designed to determine the combustibility of dry products. In this test, the ignition of a powder layer with a glowing platinum wire is attempted; the test is performed at atmospheric and at reduced pressure.

The second test consists of heating some of the powder on a hot (240–360°C) metal surface. A temperature of self-ignition is assigned if ignition or decomposition occurs within five minutes.

The same source also recommended a Through-flow Layer Ignition Test, in which air flows through an unfluidized bed while the temperature is kept either constant or gradually increased with time. The test is meaningful for fluid bed and for conveyor dryers with a maximum temperature of 450°C.

The Cross-flow Layer Ignition Test is meaningful for tray and spray dryers. It is carried out by exposing a layer of material, that is deposited on a thin metal plate in an oven, to an airflow at constant temperature. Any exothermic activity is detected by an embedded thermocouple. The test is continued until ignition has occurred or until it is evident that ignition will not take place. The results of this test can be interpreted by means of a model.

The minimum ignition temperature for bulk deposits can be determined by means of a Dewar vessel. The material is introduced into a Dewar at different initial temperatures and the sample's temperature is recorded as a function of time. To obtain a safe bulk-storage temperature, 50°C is subtracted from the minimum ignition temperature.

In general, the results of the tests can give guidance on the selection of safe operating conditions, but the flammability of any gases evolved when the material decomposes must also be taken into consideration.

Other possible methods to remove or reduce the fire hazard can be adopted, e.g., operating under vacuum (no oxygen, low temperature), inerting the atmosphere (no oxygen), and exclusion of internal agitators which potentially can provide ignition sources by mechanical friction. Bartknecht (1987) has discussed a sophisticated series of tests to detect the flammability of dusts. Normal firefighting methods should be adhered to for drying operations, e.g., provision of sprinkler systems, etc.

The flammability of solvents is discussed under vapor explosions (Section 12.4).

12.3   Dust explosions

Introduction

Dust from a combustible solid, when suspended in air, may, under certain circumstances, cause a dust explosion. Frequently, dust explosions have occurred in the past during grain milling, grain storage, and coal-mining operations and have also resulted from drying operations in the chemical industry. These explosions have often resulted in severe personnel injury and in material damage. If an explodable dust can occur anywhere in a drying system, the possibility of an explosion cannot be excluded unless preventative measures are adopted.

For example, explodable dust clouds can form in the baghouse of a spray dryer (dust collection is hazardous because it serves to collect all the fines in one place) or in a fluid-bed dryer at the end of the drying cycle (dry material has a greater propensity for explosion than wet material). Fuel, oxygen, and heat are all required for a dust explosion, and it is usually very difficult to exclude a potential source of heat in a specific drying system. The ignition sources for fires mentioned in Section 12.2 can also trigger a dust explosion in the presence of a dust cloud.

Dust and vapor explosions are similar in that they are rapidly proceeding fires which, on confinement, cause pressure increase. However, there are two principal differences:

1. The ignition sensitivity of vapor-air mixtures is normally much greater than that of dust-air suspensions.
2. Dust-air suspensions are usually heterogeneous since dust tends to settle while vapor-air mixtures are homogeneous and, for all practical purposes, do not segregate.

Primary and secondary dust explosions are known. Primary dust explosions can lead to a dispersion of secondary dust in the air, and this resuspended dust may lead to a secondary explosion. The possibility of a chain reaction is an inherent risk when dealing with combustible dusts. It is not within the scope of this section to give detailed instructions regarding the prevention of dust explosions or the protection against the effects of a dust explosion. For this information, the reader is referred to Field (1982), Bartknecht (1978), Palmer (1973), VDI 3673 (1979), National Fire Protection Association (NFPA) (1978), IChemE (1977) and Bartknecht (1987).

Qualitative aspects

It is important to determine whether or not a specific dust is explodable. Different detection tests exist in different countries, e.g., the modified

Hartmann apparatus with a hinged lid (Fig. 12.1) is used in Switzerland (Field, 1982).

The severity of any dust explosion depends on (a) the chemical composition of the material; (b) the surface nature of the material; (c) the dust concentration (dust explosions can only occur if the concentration of the dust is between the lower- and upper-explosion limits); (d) the particle-size distribution (smaller particles cause stronger explosions because the interfacial area is inversely proportional to the particle size); (e) the moisture content (water phlegmatises); (f) the strength of the ignition source (weak

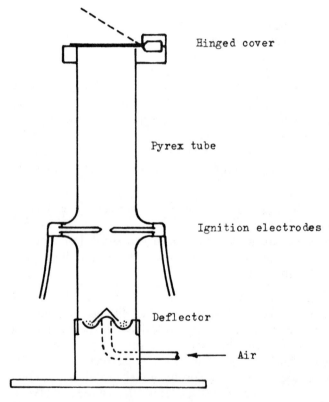

Figure 12.1 Modified Hartmann apparatus. The ignition source is either an electric spark or a heated electric coil. The dust to be tested is examined either as received or milled to below 30 microns. A pretreatment comprises drying under vacuum for an hour at 40–50°C. Dust is dispersed from the base by means of air at 7 bar to give in the 1.2–l volume dust concentrations ranging from 30–1,000 g/m^3. If a flame extends at least half the length of the tube, the dust is considered "positive."

sources can have a moderate effect and vice versa, see, e.g., Hay and Napier, 1977); (g) the oxygen concentration; (h) the degree of turbulence; and (i) the temperature and pressure.

Note that this list is not exhaustive but covers the principal factors.

Hybrid mixtures are suspensions of dust in air-solvent vapor mixtures. A synergistic effect may exist between the vapor and the dust. In these cases, it is advisable to safeguard by lowering the oxygen content to below the *maximum permissible oxygen concentration* (MPOC), i.e., safeguarding by inerting.

Quantitative aspects

If the explosibility has been shown by means of a qualitative investigation, the quantitative aspects should be investigated. The quantitative aspects are the ignitability of the dust cloud and the explosive effect.

Minimum ignition energy (MIE)

Normally, the minimum ignition energies of vapor-air mixtures are much lower than those for dust suspensions. The MIEs for gases, like hydrogen and methane, are 0.03 and 0.3 mJ, respectively. For dusts, the NFPA Guide No. 68(1978) lists MIEs ranging from 10 to 1,000 mJ. However, recent investigations have shown that the MIEs of dusts can approach those of vapors if the ignition spark's discharge time is long enough. Adjusting this reduces the "shock-wave" effect that pushes the particles away from the spark (Field, 1982).

Until recently all MIEs were determined in a Hartmann apparatus that gives ignition energies ranging from 3 mJ to 2 J. Many values can be found in the literature. A value of less than 100 mJ indicates that, in principle, it would be possible to ignite dust by means of an electrostatic discharge from personnel.

Minimum ignition temperature (MIT)

MITs are determined by passing a dust-air suspension through a laboratory furnace and determining the temperature at which ignition takes place. Typically, values of about 400°C are found, but some materials can be considerably below this level (e.g., sulphur dust has a value of 190°C). The value obtained from the test is normally higher than the "real" figure since the air-dust mixture entering the furnace is cold whereas in practice the suspension is usually preheated. Standard practice in Germany is to take two thirds of the measured MIT.

Minimum explosible concentration (MEC)

Explosible dusts have a minimum and a maximum explosible concentration; i.e., a dust explosion cannot occur when either the dust concentration is below the minimum explosible concentration or when the dust concentration exceeds the maximum explosible dust concentration. The explanation is that the progress of a deflagration is hindered because of phlegmatization by the air (at low concentration) or by the dust (at high concentration). It is normal to establish MECs in a 20-liter spherical apparatus (Figure 12.2), using a chemical igniter supplying 10,000 J to ensure ignition. Usually, MEC values are in the range of 10–60 g/m^3. Sometimes, it is possible to safeguard a system by ensuring that the concentration of the suspension is always below the MEC.

Maximum permissible oxygen concentration (MPOC)

MPOCs can be determined in a modified Hartmann apparatus, with typical values being 10% by volume. Usually, the MPOC decreases sharply with

Figure 12.2 A twenty-liter spherical explosion apparatus. (Courtesy of Kühner, Basel, Switzerland.)

an increase in temperature. It is possible to be protected against dust explosions by keeping the oxygen in the system below the MPOC; however, relief venting is more common.

Explosion effects

Both the maximum pressure (MP) and the maximum rate of pressure rise (MROPR) will be discussed in this section. The highest pressure occurring in a confined space that is due to a dust explosion is termed the MP. If the equipment exceeds certain minimum sizes, the MP is independent of the equipment size; however, the MROPR

$$\left(\frac{dp}{dt} \right)_{max}$$

is strongly dependent on the equipment size. Both variables can be conveniently determined in a 20-l spherical apparatus. A typical record is given in Figure 12.3. The MROPR so determined can be used to calculate K_{st} by means of the cube-root law:

$$\left(\frac{dp}{dt} \right)_{max} \cdot V^{1/3} = K_{st} \tag{12.1}$$

The cube-root law is applicable to small- and to large-scale equipment and can therefore be used for scaling-up. K_{st} is a constant for a given dust. Equation 12.1 states that the MROPR values measured in large pieces of equipment are smaller than those established in small apparatus. This can be explained by assuming that the linear burning velocity is constant and that therefore the combustion in smaller equipment is complete in a shorter period of time than the combustion in larger equipment. The K_{st} value so determined is used for the sizing of relief vents; the higher the rate of pressure rise, the greater the vent area required. Furthermore, the cube-root law shows that the size of the relief vent can be relatively small for large-volume equipment.

The applicability of the cube-root law is one reason for permitting the amount of vent area per cubic meter to decrease on scaling-up.

VDI 3673 (1979) described the sizing of relief vents. The research was based on tests carried out on equipment that ranged from 1 to 60 m³. The guidebook, which was produced, advised that tests should be carried out on 1-m³ scale. However, it has been shown that results from standard 20-l test spheres are in line with the 1-m³ values (Bartknecht, 1987). Standard practice is to investigate dry dusts using only materials of particle size less than 63 microns. The igniter delay is 60 msec, which implies a certain degree of turbulence. Other delays will lead to different K_{st}-values. The same ignition delay is applied to testing in both 20-l and 1-m³ apparatus.

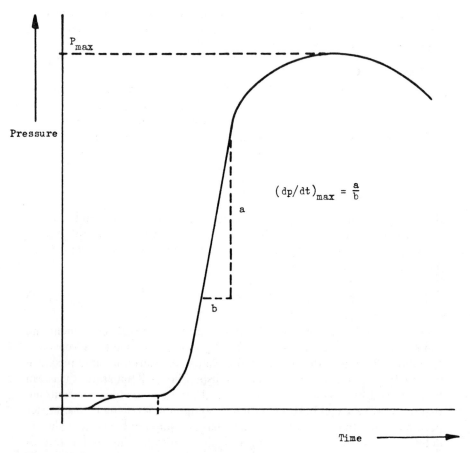

Figure 12.3 A pressure-time record of a dust explosion test.

Three classes are used for the classification of K_{st}-values:

Class	(K_{st}), bar · m · sec^{-1}
1	$K_{st} \leq 200$
2	$201 \leq K_{st} \leq 300$
3	$K_{st} > 300$

These results can be used for equipment volume of up to 1,000 m³ provided that $L/D \leq 5$ and that the equipment is capable of withstanding an overpressure of at least 0.2 bar. Relief venting is a method of safeguarding in which one section is designed to be weaker than the rest of the equipment.

This method can be used as a safeguard only if the weakened section can open rapidly (low inertia cover, less than 10 kg/m^2).

Safeguarding

As mentioned, two fundamentally different categories of safeguarding exist: one that is based on prevention and the other on curing. Since methods based on prevention are preferable, this group will be covered first.

Good housekeeping

The avoidance of unnecessary dust suspensions and accumulations will considerably reduce the possibility of a dust explosion.

Eliminating ignition sources

Another important preventive aspect is the use of the proper class of electrical hardware and wiring to avoid electrical sparks. The electrical equipment should be dust tight or nonsparking and should not produce surface temperatures capable of igniting the dust. Also, electrostatic discharges must be controlled. Efficient grounding of equipment is of paramount importance, and the installation of bonding strips across joints is essential. Antistatic materials can be used for conveyor belts, clothing, footwear, and flooring. Steel- or carbon fibers interwoven into the fabric of the filter bags of fluid-bed dryers are helpful. The entrance of stray metal on combined milling/drying must be avoided and often metal detectors are installed to check the feed. IChemE (1977) has advised that the drying air temperature be kept below the MIT of the dust (margin at least 50°C) and below the MIT of a deposit (margin at least 20°C).

Direct firing entails the risk of incandescent particles entering the dryer with the drying air. Firefly, a Swedish firm, developed an infrared radiation detector capable of reacting to glowing particles that have temperatures as low as 400°C. The detector can activate an extinguishing system (see p. 280). Originally, the system was applied to pneumatic conveying systems in the wood-processing industries of Scandinavia. Applications have also been found in starch drying. Another unnecessary ignition source is welding. A careful inspection should precede such an activity. Finally, spontaneous product heating must be considered (see Section 12.1).

Inerting

Inerting with nitrogen provides the most effective means of preventing dust explosions. It is essential to monitor the oxygen content of the gas. The main reason that this method is not applied more often is the cost. The inert gas is recycled but the drying system is under a slight overpressure to prevent oxygen ingress. The overpressure causes leakages and a nitrogen make-up is required.

Protection methods based on curing follow.

Containment

The equipment must be able to resist pressures of 7–10 bar, which is not feasible except for smaller pieces. Normally, drying equipment can withstand overpressures of, e.g., 0.2 bar. An alternative is to build the equipment "pressure-shock-resistant." An explosion can then be contained but may necessitate repairs or even replacement.

Plant separation

To prevent the spread of an explosion processing plants ought to be adequately separated. Rotary valves and screw conveyors may act as "chokes."

Relief venting

Venting is used widely as a safeguarding method. It does not prevent a dust explosion from occurring but protects the equipment against damage. This method can be used only if the emission of material is allowable. It must be ascertained that the relief vents are not prevented from opening by rust or ice. It will usually not be possible to emit the burning dust and vapor into the plant and ducts are often used. However, a careful design is required because the flow through the duct entails a pressure loss.

Sample calculation   Dust is tested and the K_{st}-value obtained indicates that it belongs to the class St 2. The result must be used to size a fluid-bed dryer's relief vent. The volume is 10 m^3. The relief vent is expected to open at a pressure of 1.1 bar (absolute) and the pressure in the dryer cannot exceed 1.2 bar.

Consult Figure 12.4, which is a reproduction of nomograph 7a of VDI 3673 (1979). Enter at 10 m^3 and move vertically up to the line reading P_{red} = 1.2 bar. Go left to the line reading St 2 and read 2 m^2 relief vent area. The graphs of VDI 3673 (1979) are valid for dusts of classes St 1 and St 2 if P_{max} ≤ 10 bar. Also, they are valid for dusts of class St 3 if 10 < P_{max} ≤ 12 bar.

Suppression

Suppression is a safeguarding method which is activated when the start of a dust explosion is detected and an extinguishing agent is distributed to dissipate the explosion. The operation of this protection is discontinuous; the safeguarding does not function during the normal operation but starts working upon a signal. Suppression requires more than one stage to achieve safety, e.g., pressure detection, conveying a signal, and the subsequent reaction. Its functioning is dependent on equipment checking and maintenance. Proper operation is not continuously apparent. An outline of this system can be seen in Figure 12.5 The signal from the detector can also be used to provide an alarm, to close valves, to shut down equipment, and so on.

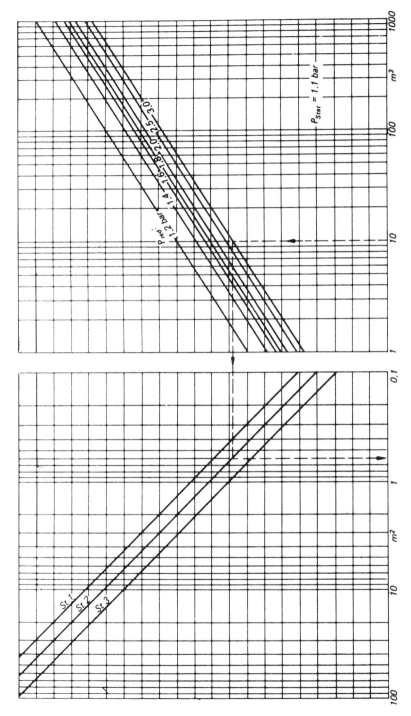

Figure 12.4 Nomograph for sizing relief vents. Relief vent opens at an absolute pressure of 1.1 bar. (Courtesy of VDI, Düsseldorf, Federal Republic of Germany.)

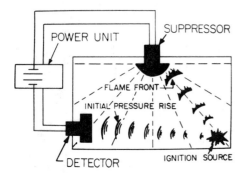

Figure 12.5 Simplified suppressor outline. (Courtesy of Fenwal, Ashland, Massachusetts.)

The development of the suppression system was inspired by experiences during World War II. At that time the Royal Air Force was suffering severe losses of combat aircraft. More than 50% of the losses caused by gunfire were the consequence of fuel-tank explosions; structural damage from bullets had relatively little effect on the planes.

The RAF started to investigate whether tank explosions could be stopped after they had started, and their investigations were successful. However, with the arrival of surface-to-air and air-to-air missiles, the protection system became obsolete. A hit by a missile meant destruction of the aircraft regardless of the fuel-tank protection (Martin, 1978). Then the applicability of this safeguarding method was checked for the process industry. The first field installations appeared during the 1950s: coal dust, starch, plastics, and pharmaceuticals. Whether this protection method was fast enough might give cause to wonder.

A sample calculation illustrates the feasibility of this approach. A dust explosion starts at a 2-m distance from a detector. The signal of the pressure rise travels at the speed of sound to the detector (342 m/sec) and therefore detection time is approximately 6 msec. For all practical purposes, the electrical actuation is instantaneous. A typical dispersal speed of the suppressing agent is 60 m/sec. The suppressor is located at a 2-m distance from the developing fireball. The suppressing agent requires about 34 msec to arrive. So, 40 msec elapse between detection and dissipation. Now, the question is: How far has the dust explosion advanced? A typical flame speed is 3 m/sec. Remember that dust explosions start out with relatively low linear flame speeds. If they are allowed to progress, the flame speeds would accelerate sharply. However, we are dealing with the first stage, and, consequently, the dust explosion has advanced 12 cm in 40 msec.

This sample calculation clearly illustrates that the protection method can be quite adequate. However, detection should be right at the beginning and it must be possible for the flow of suppressant to reach the ignition source immediately.

An important advantage of this protection method is that a dust explosion is not followed by an emission, which is particularly attractive when dealing with toxic substances. The example elucidates that a radiation detection (see the previous remark on Firefly, p. 279) hardly offers a bonus over a pressure detection. Basically, both the speed of sound (342 m/sec) and the speed of light (300,000 km/sec) exceed the speed of flame (3 m/sec) considerably.

Well-known suppressing agents are monammonium phosphate (a powder) and the halons (liquids). Halons are halogenated hydrocarbons with the compounds having code names, such as Halon 1011: the first figure indicates the number of carbon atoms per molecule, whereas the second, third, and fourth give, respectively, the number of fluorine, chlorine, and bromine atoms present in the molecule (e.g., Halon 1011 is CH_2ClBr). The function of the halons is to inhibit reactions that are essential for combustion.

An attractive aspect of the halons is that they do not contaminate the process as they evaporate. Halons are more rapidly dispersed than monammonium phosphate. Practically, the halons can be discharged from hemispherical suppressors that are present in the drying system. The release can even be by means of chemical detonators. Cases can be met where the hemispherical suppressors would be damaged by the process (e.g., in milling operations). High-rate discharge bottles with 20-mm single or double outlets or 75-mm outlets can then be used. These discharge less fast than hemispherical suppressors but are less sensitive and have higher capacities. A typical high-rate discharge bottle is pressurized with N_2 (20 bar).

A limitation of the technique is the path the suppressants can be expected to travel, with possibly a maximum value of 4 m (Field, 1982). Moore (1986) reported successful testing on a 250-m^3 scale. Graviner (1987) stated that, as a general summary, their explosion-suppression systems are effective against most solvent vapor explosions, hybrid-mixture explosions, fuel-droplet explosions, and dust explosions provided that the explosibility rate constant

$$K_g, K_{st} < 300 \text{ bar} \cdot \text{m} \cdot \text{sec}^{-1}.$$

For more violent materials and for weak components that are located within a building such as larger spray dryer complexes, the combination of explosion venting and explosion suppression is an alternative.

Suppression has found applications at, e.g., the flash drying of starch (Meinhold and Carney, 1967). A further application is during the combined

Figure 12.6 A suppressor installation. (Courtesy of Graviner, Colnbrook, Slough, England.)

Figure 12.7 A membrane pressure detector. (Courtesy of Graviner, Colnbrook, Slough, England.)

Figure 12.8 A high-rate discharge explosion suppressor. (Courtesy of Graviner, Colnbrook, Slough, England.)

milling/drying of explosible materials. Figure 12.6 exhibits an installed suppression system, Figure 12.7 depicts a membrane pressure detector, and Figure 12.8 shows a high-rate discharge explosion suppressor with a capacity of 5.4 liters.

12.4. Vapor explosions

It is assumed that vapor explosions in drying equipment are deflagrations and can occur only if the vapor concentration in the air is within the explosive range. The lower explosion limit for most organic solvents is of the order of 1 to 2% by volume, with methanol's being an exception: 6.7% by volume. Safetywise, it is possible, in principle, to use air to evaporate the combustible solvents; however, keeping out of the explosive range is then essential. If the safeguarding is based on an operation below the lower explosion limit (LEL), the actual concentration should be well below the

LEL. The recommendation is not to exceed 25% of the LEL in the exhaust air duct at any time (IChemE, 1977) (see the section on Qualitative Aspects that also treats hybrid mixtures, p. 275). The flash point of a solvent is the lowest temperature at which a solvent vapor-air mixture in equilibrium with the liquid can be ignited. So, when material wetted with an organic solvent is stored in a hopper at a temperature above the flash point, the vapor head is explosible.

Acetone's flash point is $-18°C$ and methanol's is $12°C$. The auto-ignition temperature of a solvent is the lowest temperature at which an explosible vapor-air mixture can be ignited by a hot surface in standard apparatus. For example: most auto-ignition temperatures are above $200°C$, for methanol it is $470°C$, and for acetone $540°C$. In principle, the cube-root law is valid here too:

$$\left(\frac{dp}{dt} \right)_{max} \cdot V^{1/3} = K_g \qquad (12.2)$$

K_g is for solvent vapors in the range 40–80 bar $\cdot$ m $\cdot$ sec^{-1} on measuring in the 20-l apparatus. However, this is with an ignition energy of only 10 J. It is known that higher K_g-values can be found when the ignitor supplies more energy. Compare also the 10-J value with 10,000 J as normally used for dust explosions. Higher values can also be measured in larger equipment (up to ten times higher) because of increased turbulence (Moore, 1984 and Bartknecht, 1978). The assessment of K_{st}-values for dusts has a greater practical significance than the determination of K_g-values for gases. The background is that the K_{st} and the P_{max} for a given dust can be used to establish vent sizes. Usually, this is not done for the protection of dryer operations on dealing with explosible vapor-air mixtures because the minimum ignition energy in these cases is quite low. So, the ignition of a vapor-air mixture is more probable than the ignition of a dust-air mixture. A second reason is that K_g is not really a constant for a given vapor; much higher values can be found in larger than in smaller apparatus (even when keeping the ignition delay constant). Moore (1984) attributed this finding to acoustically coupled accelerations set up during the course of quiescent gas explosions in large volumes (e.g., 10 and 25 m^3). The usual protection method is inerting.

12.5 Additional remarks

General

A distinction between convective dryers and contact dryers can be made. Both types can be equipped with safeguarding systems. Some can be provided with additional options; e.g., the pressure detector that initiates a suppression can, at the same time, (a) activate a detonator-operated bursting disk, (b) introduce a suppressant into an area of the plant that is away from the place where the explosion started to prevent a second explosion or fire (advance inerting), (c) isolate one area of a plant from another by means of explosion isolation valves, (d) flood hazardous areas (deluging), or (e) shut down the drying plant.

The aforementioned provisions require activation in order to be effective. An option that prevents the passage of flames from one part of a drying system into a different one and that does not need activation is the flame arrestor, which is usually mounted in a line.

Convective dryers

For most dryers of this type it is possible, in principle, that an explosible dust (if present) is ignited by the drying air. IChemE (1977) therefore recommends for a number of dryers to choose the inlet air temperature at least 50°C below the minimum ignition temperature for the dust. In this respect, the fluid-bed dryer is an exception because the drying gas must pass the product layer and thereby cools down. Deposits may occur in convective dryers, their location depending on the type of dryer; e.g., the top bend of a flash dryer's tube is vulnerable. A further example is a fluid-bed dryer's distribution plate, which is virtually at the air-inlet temperature. The vulnerability of deposits to ignition can put restrictions on the air-inlet temperature or exit air temperature. The material leaving the dryer must be cool enough to avoid bulk ignition.

Dealing with solvents almost automatically calls for inerting. An inert gas recycle with solvent condensation is necessary. Contact dryers have an edge over convective dryers with regard to the evaporation of solvents, since much less gas is involved. Venting is a widely used safeguarding method if the feed is water wet.

Fluid-bed dryers

Bartknecht (1978) discussed explosions that have occurred in batch dryers on evaporating solvents. Inerting is probably the protection method to adopt in this case. Because of their very nature, fluid-bed dryers are generators of static electricity, and bonding is essential here. Good conditions have been obtained by incorporating conducting fibers into the fabric of filter bags.

Bartknecht (1978) also discussed the fact that less venting area is required than prescribed by VDI 3673 (1979). The background is that a fluid-bed dryer, in a way, deviates from the cubic form ($L = D$).

Rotary dryers

A well-known example of a fire risk is the drying of sugar beet pulp with air of about 800°C. Normally, the pulp does not reach a temperature at which ignition occurs. However, if the process stops because of a failure of the electricity supply, the pulp may ignite because it is then being heated up by the hot shell. Because the shell is rotating, it is difficult to locate parts of a protection system on it; therefore, relief or suppression must be arranged on or from the heads.

Flash dryers

Meinhold and Carney (1967) described the safeguarding of starch drying by means of suppression.

In one case it appeared necessary to raise the product's temperature fairly high in order to reach the desired low final moisture content. However, this temperature was close to the ignition temperature. Consequently, inerting was chosen, and the drying gas was recycled and its O_2-content monitored.

Spray dryers

Often, relief venting is used as a protection method. As reported in VDI 3673 (1979), this practice leads to relatively large areas because of the large volumes. Thus, it is a common practice to reduce the obtained areas because the full volume of a spray dryer is not filled with an explodable mixture.

Combined milling/drying

Bartknecht (1978) described how difficult it is for a dust explosion to develop itself fully in a mill. The reason is that mill internals deviate strongly from the cubic form. A critical risk is the entrance of tramp metal, despite the fact that a metal detector does check the feed. Many mills are protected by means of suppression.

Baghouses

Many convective dryers are equipped with baghouses. Frequently, dust can accumulate in equipment and the potential for a dust explosion or a fire can be considerable. The contents can be ignited by means of a flame front entering from the dryer through the line. Relief venting is a common protection method.

A series of incidents occurred in the wood industry, probably initiated by glowing dust particles. The standard practice is now to locate the filter at a safe distance and to apply relief venting.

The location at a safe distance (e.g., in the yard), is not always possible in the chemical industry. However, it is important to realize that the yielding of the vent relief emits much burning material which should not be allowed to create a harmful situation.

Contact dryers

Generally speaking, contact dryers can be safeguarded easier than convective dryers. Inerting is not too difficult because the strip gas flow is much smaller than the gas flow on convective drying. Contact dryers can often be built strong enough to be able to contain an explosion.

However, there are aspects that must be observed closely. The product acquires the wall temperature and this must not cause ignition. Often, contact dryers are equipped with an agitator, and this feature entails the risks of the occurrence of hot spots (bearings), frictional heat or sparks that are due to tramp metal or a slight shaft misalignment. Furthermore, the dust concentration is often within the explodable limits. Pfaudler tumblers or the like are vulnerable to the generation of static electricity. It must be realized that dust explosions can also occur in vacuum dryers, but rarely at pressures below 50 mbar.

Relief venting and suppression can be practiced; however, there are specific ins and outs. To give an example: both relief venting and suppression are not adequate for plate dryers because of their relatively complicated construction. Inerting is the answer in this instance.

Notation

D	Diameter	m
K_g	Vapor explosion constant	bar · m/sec
K_{st}	Dust explosion constant	bar · m/sec
L	Length	m
P_{max}	Maximum explosion pressure (gauge)	bar
P_{red}	Maximum pressure on venting	bar
P_{stat}	Pressure at which relief vent opens	bar
p	Pressure	bar
t	Time	sec
V	Volume	m³

References

Bartknecht, W. (1978). *Explosions: Course and Protective Measures,* Springer-Verlag, Berlin (in German).

Bartknecht, W. (1987). *Dust Explosions: Course and Protective Measures,* Springer-Verlag, Berlin (in German).

Field, P. (1982). *Dust Explosions*, Elsevier, Amsterdam.

Gerritsen, H. G. and C. M. van 't Land (1985). Intrinsic continuous process safeguarding. *I & EC Process Design & Development*, 24, 893.

Graviner (1987). Private communication.

Hay, D. M. and D. H. Napier (1977). Minimum ignition energy of dust suspensions, *IChemE Symposium Series No. 49*, IChemE, Rugby, England

IChemE (1977). *User Guide to Fire and Explosion Hazards in the Drying of Particulate Materials*, IChemE, Rugby, England.

Martin, A. J. (1978). Explosions: The pause that suppresses, *Waste Age*, 9, 36.

Meinhold, T. F. and R. D. Carney (1967). Over 300 kinds of starch processed in one dryer, *Chem. Proc.* (Chicago), 30.

Moore, P. E. (1984). Explosion suppression trials, *The Chem. Eng.*, December (409), 23.

Moore, P.E. (1986). Towards large volume explosion suppression systems. Are they too expensive and can they cope? *The Chem. Eng.*, November (430), 43.

National Fire Protection Association (NFPA) (1978). *Guide for Explosion Venting NFPA No. 68*, NFPA, Boston.

Palmer, K. N. (1973). *Dust Explosions and Fires*, Chapman & Hall, London.

VDI (1979). *Pressure release of dust explosions, VDI No. 3673*, VDI-Verlag, Düsseldorf (in German).

13

CONTINUOUS MOISTURE-MEASUREMENT METHODS, DRYER PROCESS CONTROL, AND ENERGY RECOVERY

13.1   Introduction

Four different aspects are treated in this chapter: (a) continuous solids moisture-measurement methods, (b) continuous gas moisture-measurement methods, (c) dryer process control, and (d) energy recovery from exiting flows.

Measuring the moisture content of the feed or the product can be useful for process control; the same reasoning can be applied to the measurement of the gas's moisture content. However, in many instances, practical dryer process control methods are based on temperature measurements. The large energy consumption of dryers per se and the energy prices often call for an energy recovery. Practical recovery methods are restricted to recovery from gaseous flows.

13.2   Continuous solids moisture-measurement methods

Introduction

It is possible to choose two different criteria to classify moisture-measurement methods of solids. First, a distinction can be made between batchwise functioning and continuous systems. Second, absolute methods can be compared to inferential methods. In absolute methods, water is extracted from the material by oven drying, desiccation, distillation, titration, or by reaction (e.g., Karl Fischer method). They do not need calibration. Some physical property of the material under consideration, which varies with the quantity of water present, is measured in inferential methods. They do need calibration. The degree of binding of the water molecules to the host material affects the results of most inferential methods of moisture measurement. Hence, the calibration results for one material do not apply to a different one.

Absolute methods are not discussed in this section because invariably they are carried out batchwise and only continuous methods are dealt with. The inferential methods can be practiced in continuous and in batchwise operations. For example, the infrared reflection of a static sample and a product on a moving conveyor belt can be measured. The continuous solids moisture-measurement methods discussed are: (a) electrical conductance, (b) electrical capacitance, (c) infrared reflection, (d) microwaves, (e) neutron moderation, (f) nuclear magnetic resonance, and (g) temperature.

When selecting a method, a feasibility study is carried out first. The moisture content of a material of high electrical conductivity cannot be measured with a capacitance technique. A narrow particle-size distribution combined with a small average particle size is more favorable for infrared reflection than a wide distribution combined with a large average particle size (e.g., larger than 1 mm). Neutron moderation and nuclear magnetic resonance concentrate on all the hydrogen atoms present and not only on the hydrogen atoms in water. For obvious reasons, neutron moderation requires safety precautions.

The second step is a laboratory assessment of available measurement techniques, with the successful methods identified. On-line calibration trials come next and, finally, a system is commissioned.

Electrical conductance

Measuring the electrical conductance to assess the water content is often applied in the textile industry. First, it is important to distinguish between materials like wool, cotton, and rayon and those formed of synthetic fibers (e.g., polyester and nylon). The former class of materials contains in its ready-for-use form much more water than the latter class; wool, e.g., contains up to 35% water by weight and nylon approximately 5% by weight.

A consequence of this difference is that electrostatic charges, during the drying of moving synthetic materials, can interfere with the conductivity measurement. The greater water content of the materials in the other class enables the electrostatic charges to be removed by means of bonding. So, continuously moving cellular materials can be checked by means of conductivity measurements while moving synthetics cannot. However, both cellular and synthetic textile materials can be checked by means of conductivity if the products are stationary.

Second, note that it is possible to differentiate between bound water contents only. The instruments lack sensitivity to distinguish between water contents when upon touching the material is clearly wet. Cotton conductivities at 3.5% and 11.5% water by wt differ by a factor of 10^5. This great difference, that is due to the different water contents, by far exceeds differences that are due to other variables, such as the mass of the material in g/m^2 and the cloth thickness.

The method is also applicable to wood, tobacco, and cereals. Here, too, the conductivity varies markedly when the bound water content varies. The result of the measurement is usually only slightly dependent on the temperature. Contact, twin-needle, multiple-needle, and comb electrodes exist for various applications. The moisture must be uniformly distributed in the material, otherwise the current takes the shortest possible path between the electrodes. The moisture content of fibers is determined by means of electrodes that are built as rolls (see Fig. 13.1). Alternating current is used to avoid polarization. Figure 13.2 depicts an instrument for rack-mounting.

Figure 13.1 Continuous measurement of the moisture content of sheet material by means of conductivity. Three electrically insulated rolls are connected to the amplifier. Measurement is made of the potential drop that is due to the cloth's resistance. (Courtesy of Mahlo GmbH & Co KG, Saal/Donau, Federal Republic of Germany.)

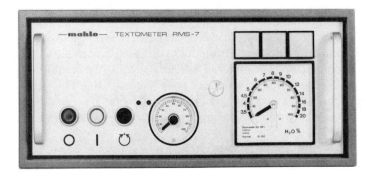

Figure 13.2   An instrument for measuring moisture contents by means of conductivity. (Courtesy of Mahlo GmbH + Co KG, Saal/Donau, Federal Republic of Germany.)

Electrical capacitance

Introduction

The method relies on a change of the capacitance of a capacitor when the water level of the process flow changes. Generally, the capacitance depends on three variables: the area of the electrodes, the distance between the electrodes, and the dielectric constant of the material between the plates. At ambient temperature and for AC frequencies smaller than 10 Hz, the dielectric constant of water is approximately 80. The dielectric constant (DC) of most other materials is in the range 2–8. So, the water level influences the overall DC markedly.

The method is unsuitable for materials that can conduct electricity. In many instances, the material to be analyzed moves over a probe's surface. However, a direct physical contact is not always necessary; e.g., a measuring head can be located under a moving conveyor belt. The distance between the head and the material shall not exceed several millimeters and must remain constant.

Background

The capacitance of a capacitor is directly proportional to the dielectric constant of the material between the plates. Some DC-values:

Water: 80
Methanol: 33.6
Fats, oils: 3 (approximately)
PVC: 3
Teflon: 2.2

Sulfur: 4
Glasses: 5.5–7
Ice: 3.1

The explanation for water's anomalous behavior is that water molecules are dipoles that, on application of an electric field, align with the field that is thus reinforced. It is interesting to compare water and ice. Apparently, the ice molecules cannot orient themselves. This fact explains the phenomenon that the on-line moisture measurement based on capacitance is less effective for bound water than for free water.

The low DC-values of several solids can be noticed. The (small) conductivity of the material can, in principle, disturb the result to some degree. By employing sophisticated electronics, it is possible to eliminate this effect completely. In practice, the effect can also be removed by using frequencies exceeding 10^6 Hz. This frequency, which is used by Brabender (FRG), is the industrial frequency 13.56×10^6 Hz. An advantage of this method is that it presents an overall picture, i.e., local water level fluctuations do not interfere. The usual practice is to have the electrodes in line, as depicted in Figure 13.3. The field is indicated schematically by the curved lines. An important advantage of this set-up is that the instrument's reading does not depend on the layer thickness of the material. However, a minimum layer depth of approximately 10 mm must be observed (Fig. 13.4). The technique has found many applications for agricultural products (coffee, grain) and for such materials as sand and fertilizers. If the bulk density of the analyzed material varies, it is recommended that the variable be measured simultaneously by means of a radio-active source. The combination of the two signals may still give a good indication of the moisture content.

Practical aspects

Variables that influence the instrument's output include the water content, the way water is held by the material, and the bulk (packing) density. (Some of these variables have already been discussed.) Further aspects are: (a) the temperature (effect on material properties), (b) the dry material's DC, and (c) the presence of impurities (small amounts of impurities can have pronounced effects).

Infrared reflection

Introduction

The method is based on the ability of water to absorb infrared light. The light is cast upon the material to be analyzed and part of it is reflected. The

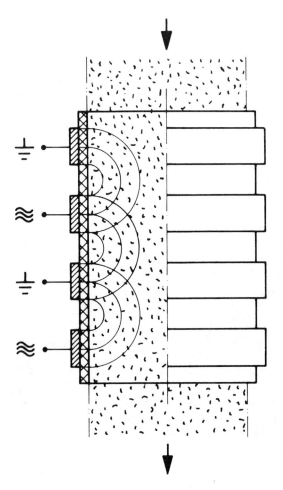

Figure 13.3 A schematic of a continuous on-line DC moisture measurement cell. (Courtesy of Brabender Messtechnik KG, Duisburg, Federal Republic of Germany.)

reflected light's intensity is a measure of the water content. Infrared reflection is a contactless method that gives information only of the material's surface. It is substantially independent from the bulk density (see Fig. 13.5).

The method is to be distinguished from infrared transmission. There, the emission source and the detector are located at different sides of the material. However, IR transmission also relies on the absorption of radiation and can be applied to web and sheet material.

Figure 13.4 Continuous on-line DC moisture measurement cell. (Courtesy of Brabender Messtechnik KG, Duisburg, Federal Republic of Germany.)

Figure 13.5 An installed continuous moisture measurement system employing IR reflection. (Courtesy of Moisture Systems Corp., Hopkinton, Massachusetts.)

Background

The IR light that is used has wavelengths between 1 and 3 μm (near infrared). Infrared light can interact with the vibration of the O-H bonds in the water molecule. However, interactions with O-H bonds in other molecules (e.g., alcohols and sugars) are also possible. This means that molecules different from water should not be present or, when present, should cause a constant absorption of energy.

Frequently, the wavelengths employed are 1.45, 1.94 and 2.95 μm, with 1.45 μm and 1.94 μm being used more often than 2.95 μm. Of the two, the 1.94-μm radiation is the more powerful one. Hence, a wavelength of 1.45 μm is often used for materials that have a high moisture content (e.g., paper and textiles containing 80% water by wt). To cancel the influence of variables other than the water level (e.g., the color of the material), a reference wavelength is chosen near to the water absorption band (the word *band* is used because the NIR spectrum of free water exhibits rather broad maxima caused by the influence of H bonding). The usual set-up contains a scanning disk with the two interference filters (e.g., 1.94 and 1.8 μm). These select sequentially the chosen narrow bands from the tungsten radiation source. A lead sulfide photoelectric cell serves as detector.

The installation of a combined source/detector over a conveying belt is the usual method. Often, special provisions (plough, comb) ensure that a fresh surface is exposed. Calibration of the instrument occurs by means of a comparison of the instrument's reading with the result of an absolute water analysis (e.g., Karl Fischer). Regular checking of the readings in the same way is advisable.

Practical aspects

Variables that influence an instrument's output are:

1. *Temperature:* Cooling the source/detector is advisable when the instrument would otherwise be exposed to temperatures exceeding 40°C.

2. *Vibrations:* These might adversely affect the functioning of the instrument.

3. *Light intensity:* An instrument compensating the effect of the light intensity decreasing with time will be discussed in the section treating a measuring system; p. 299.

4. *Optics:* See No. 3, "light intensity."

5. *Surface structure:* A bar mounted over the conveyor can often assist to smooth the surface. This remedy, however, cannot compensate for intrinsic surface differences.

6. *Particle-size distribution:* A narrow particle-size distribution combined with a small average particle size is more favorable than a wide distribution combined with a large average particle size (e.g., larger than 1 mm).

7. *Chemical composition:* Small amounts of impurities or additives, e.g., 1% by wt, can influence the reading markedly.

8. *Color:* An aspect to be kept in mind, although its influence is generally ruled out by means of the reference beam. Black materials need extra attention.

9. *Distance head/bed:* Generally, the distance should be kept constant. Altering the distance by several cm can influence the instrument's output by as much as 1% by wt.

10. *Daylight:* A changing environment might influence the performance. A cylinder through which the light is passed onto the material can be a remedy.

A measuring system

A typical measuring set-up is shown schematically in Figure 13.6. The signal from the reference wavelength and the signal from the measuring

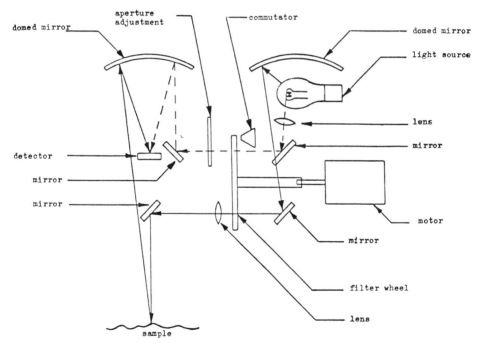

Notes:
1. Solid arrowed lines – Main Channel
2. Dotted arrowed lines – Prime Channel

Figure 13.6 Layout of the optical system of an IR-reflection instrument. (Courtesy of Moisture Systems Corp., Hopkinton, Massachusetts.)

wavelength are combined to form a ratio (reference is divided by measure). This ratio is directly proportional to the moisture level. The QUADRA-BEAM Analyzer uses two additional beams to stabilize the instrument and to eliminate drift. (Drift is the change in an instrument reading caused by factors other than moisture.) The beams have the same wavelengths as the measure and the reference wavelengths and are passed directly to the detector. Their path is indicated by means of dotted lines.

The four signals are then put through a calculation:

$$\frac{R}{M} \times \frac{M'}{R'}$$

The R and M signals are reflected from the sample. Here, a change in light intensity at the measuring wavelength will affect both the M and M' value proportionally.

Microwaves

Introduction

Microwaves are electromagnetic waves like radio waves or light. The length of the microwave varies from 1 to 10 cm and the frequency varies correspondingly from approximately 30 to 3 GHz. The idea that micro-waves could play a role in the measurement of moisture was inspired by the observation that rain attenuates radar signals.

Two independent physical properties can be utilized for measuring the moisture content by means of microwaves: the material's potential to cause dielectric losses and the material's dielectric constant. Microwaves can be used in three different ways: in transmission, by resonance, and in reflection.

Almost all commercially available instruments measure dielectric losses (absorption) by means of transmission. However, measuring the dielectric constant can offer advantages (Klein, 1987). The measurement of absorp-tion by means of transmission is contactless. Resonance cavities can be employed for foils and threads.

Background

Water molecules rotate as a whole and the energy associated with this rotation is quantized. Microwaves can interact with this movement and thus cause transitions between defined energy levels.

Since the mobility of bound water molecules is restricted, the method is more effective for free water than for bound water. Commonly used micro-wave frequencies are the S band (2.6–3.95 GHz) and the X band (8.2–12.4 GHz). The result of the measurement is strongly dependent on the bulk density and the temperature. Therefore, process instruments measure the packing density and the temperature simultaneously. The three signals are put through a calculation. The bulk density measurement proceeds by means of a radioactive source. The relationship is:

$$I = I_0 \times \exp\left(-\mu \cdot \rho \cdot D\right)$$

where

I = signal at the detector

I_0 = signal at the detector with air or water in the chamber (reference)

μ = radiation absorption coefficient

ρ = bulk density

D = line diameter

Water's absorption potential increases with increasing temperature. The presence of electrolytes affects the measurement less than it does in the capacitance technique. The size of the particulate material should not exceed one quarter of the wavelength of the microwaves to prevent undue scattering of the incident beam.

Practical aspects

Microwave instruments are applied extensively in the field of food and feed: animal feeds, dried pet foods, cereals, grain, and nuts. Here, too much moisture adversely affects the shelf life and the nutritional value. Too little moisture may make the product less enjoyable and perhaps even lead to valuable nutrient loss that is removed with the water. The right moisture content must be adhered to, and this justifies a sophisticated instrument. The method is also used for paper.

Figure 13.7 exhibits the flow chute. The walls are lined with special microwave energy absorbent material to avoid microwave reflections in the chute. Because of leakage, the sample cross-sectional area must considerably exceed the area through which the beam is sent. Figure 13.8 shows an installed system.

Neutron Moderation

Background

The method comprises the emission of fast neutrons, the conversion of those neutrons into relatively slow (thermal) neutrons by collisions with hydrogen atoms and the detection of the latter type of neutrons (see Fig. 13.9).

Neutrons are uncharged particles present in the nuclei of most atoms together with protons. The number of electrons of any atom equals the

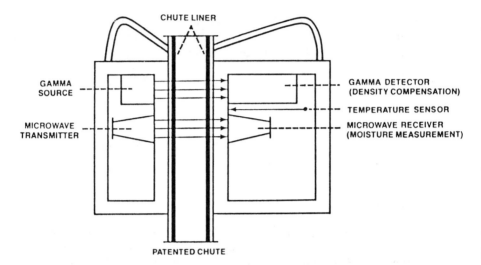

Figure 13.7 On-line, noncontacting moisture measurement by means of micro-waves. (Courtesy of Kay-Ray Inc., Wheeling, Illinois.)

number of protons in the nucleus (electric neutrality). Some radioactive sources emit fast neutrons. ^{241}Am-Be is a convenient source because it does not emit simultaneously penetrating γ-radiation like ^{226}Ra. The path the fast neutrons take is neither influenced by electric charges of electrons (negative) nor by charges of nuclei (positive). They can and do interact with the masses of the nuclei. Because of the character of the source (intensity, density), the only possible interaction mechanism is *elastic collision*. The energy lost by a neutron on colliding with a proton (a positively charged hydrogen atom) is much greater than on colliding with heavier atoms. This results since the mass of a proton and a neutron are identical for all practical purposes.

The energy loss is defined by the equation applicable for the elastic collision between two masses:

$$E_{loss} = E \times \frac{2 \times M}{(1 + M)^2}$$

where

E_{loss} = energy loss
E = energy before the collision
M = atomic mass ratio

Figure 13.8 Continuous solids moisture measurement by means of microwaves. (Courtesy of Kay-Ray Inc., Wheeling, Illinois.)

A collision with a proton causes a loss of 50%, whereas a collision with an oxygen atom leads to the loss of 11.1%.

A fast neutron becomes a thermal neutron after 15 to 20 collisions with a proton. The average free-path length required for this conversion is up to 20 cm in materials that have a high hydrogen content. The thermal neutrons are detected by a lithium iodide scintillation crystal and a photomultiplier. High water contents give rise to dense thermal neutron clouds in the detector's vicinity; thus, a signal related to the water content is being obtained. The neutron itself is also radioactive and decays with $\tau_{1/2} = 10.8$ min. It emits a β particle and a proton.

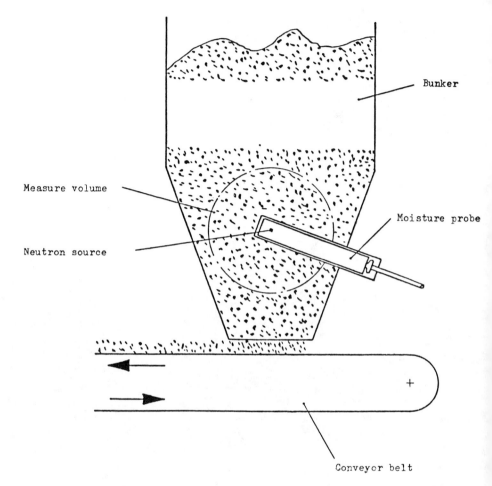

Figure 13.9 Continuous moisture measurement by means of neutron moderation. (Courtesy of Laboratorium Prof. Dr. Berthold, Wildbad im Schwarzwald,
Federal Republic of Germany.)

Practical aspects

The measurement is not influenced by temperature, pressure, pH, particle-
size distribution, or the conductivity of the sample. For each different
application, a calibration is required. Chlorine, e.g., absorbs neutrons to
some degree. The result of the measurement is also dependent on the bulk
density. When this varies, the bulk density can be measured by means of a
γ source and the detector already in use for the moisture measurement.

The neutron moderation method averages over a rather large volume. The range of the neutrons is approximately 48 cm at 25% water by volume and approximately 140 cm at 1% by volume.

Bunker probes and surface probes are known. Bunker probes do have a contact with the process material. Surface probes can be fixed on container walls and do not disturb the flux of material. They are less sensitive than bunker probes (see Fig. 13.10).

Neutron moderation finds its main application for the determination of the moisture content of heterogeneous mixes of materials, such as coke, sand, aggregates, and soil. It is not suitable for organic materials since 1% hydrogen by wt corresponds to 9% water by wt.

Safety aspects

1. If the radioactive source can fall into the process flow the process must be analyzed regarding the risks.
2. A permit to use radioactive sources is usually required.
3. A leak check must be carried out regularly.
4. Heavy shielding is required if the control range is accessible; 65 cm is a typical control range.

Nuclear magnetic resonance (NMR)

This analytical method is seldom applied to drying. Within Akzo processing plants, it has been successfully applied for the measurement of the moisture content of polyester fibers. Low-resolution NMR techniques make use of the fact that the hydrogen nucleus (the proton) possesses a

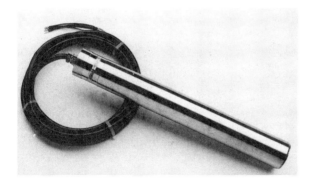

Figure 13.10 Probe for neutron moderation measurements. (Courtesy of Laboratorium Prof. Dr. Berthold, Wildbad im Schwarzwald, Federal Republic of Germany.)

spin and a magnetic moment. So, the method is specific to hydrogen and to water if water is the only component in the mixture that contains hydrogen.

A coil around the sample generates an oscillating magnetic field. The frequency is in the range of the radio waves. Coil and sample are placed in a strong and permanent magnetic field. The direction of the latter field is at an angle of 90° to the oscillating field.

The strength of the steady field is varied and at a certain set (frequency of the RF waves, steady field strength) energy is absorbed from the RF coil. This phenomenon is detected and the signal obtained is related to the moisture content.

Temperature

Temperature measurements can be used to control the moisture content of the dried product, the attraction being that temperature-measuring systems are reliable, cheap, and hardly require maintenance. Consider a batch fluid-bed dryer. This dryer is typical for the group of convective dryers. Free water (surface water) is evaporated in the first drying phase. The drying-gas temperature drops to the adiabatic saturation temperature as long as the product's moisture content exceeds the critical moisture content. The product temperature equals the adiabatic saturation temperature in that phase, which leads immediately to an important conclusion: temperature measurements cannot be used to establish moisture contents greater than the critical moisture content. Assume that drying proceeds beyond the critical moisture content. The exit-gas temperature (equal to the product temperature) starts to rise. It was established empirically that in many instances there exists a relationship between the difference (actual exit-gas temperature − adiabatic saturation temperature) and the moisture content. So, the adiabatic temperature booked in phase No. 1 can be stored electronically and used as a reference point to terminate the drying in the second phase. This principle can be simplified in many applications:

1. The continuous fluid-bed drying of vacuum pan salt is controlled by means of the highest product/gas temperature.
2. The flash drying of sodium sulfate is controlled by the exit-gas temperature.
3. The spray drying of rubber accelerators is checked by means of the exit-gas temperature.

Simplification is not always possible. Textile drying can be controlled by the difference between the indication of a light thermocouple riding on the dried material and the reading of a thermocouple measuring the adiabatic saturation temperature.

It is possible in most instances to use this principle to control the final moisture content during convective drying. The method is less suitable for performance control in contact dryers.

13.3   Continuous moisture measurement in gases

Introduction

Sometimes it is necessary to monitor the moisture content of the exhaust gas of a dryer. A high moisture content can reflect a satisfactory drying efficiency and vice versa. Furthermore, the product quality may call for a certain humidity of the gas the product was in contact with. There is a great variety of hygrometers on the market. The heart of the system is the sensor. A complication is that the exhaust gases of a dryer are usually contaminated to some degree. It is possible to use sampling systems filtering the gas and, if necessary, cooling the gas. Note that these two steps do not change the gas moisture content. On the other hand, a gas pretreatment is never perfect and complicates the measuring system.

Three different methods can be used satisfactorily in drying systems: (a) chilled-mirror hygrometer, (b) spectroscopy, and (c) dry- and wet-bulb thermometers. Dry- and wet-bulb thermometers measure the temperature depression of a wetted wick that is due to latent heat of evaporation. The depression is a measure for the water content, a method that is fairly simple. The other two methods, which are discussed in some detail, exhibit excellent long-term stability. Both the chilled-mirror hygrometer and the dry- and wet-bulb thermometers are absolute methods that hardly require calibration. Spectroscopy is inferential and does require calibration.

Chilled-mirror hygrometer

A chilled-mirror hygrometer provides continuous dew-point measurements (see Fig. 13.11). The gas flow to be analyzed flows from the left to the right through the instrument. The heart of the instrument is a mirror cooled by a heat pump. When the mirror has the wet-bulb temperature, a dew forms on the surface and the mirror's light-reflecting capability changes. This can be detected by a photoelectric cell receiving the light from a light-emitting diode via the mirror. A separate light-emitting diode/photodetector pair provides a reference light measurement. The mirror surface temperature is automatically and continuously controlled at the dew- (or frost-) point temperature of the sample gas. A precision platinum resistance thermometer, which is embedded just beneath the mirror surface, is used to accurately measure the mirror temperature.

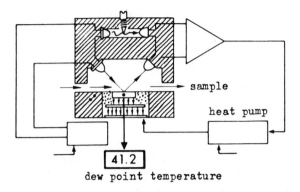

Figure 13.11 A condensation dew-point hygrometer. (Courtesy of General Eastern, Watertown, Massachusetts.)

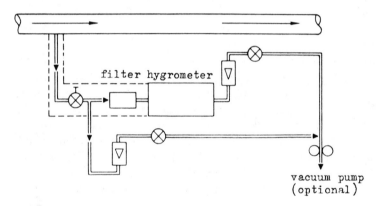

Figure 13.12 A hygrometer sampling system. (Courtesy of General Eastern, Watertown, Massachusetts.)

A typical operating range is −40 to +100°C. The instrument responds quickly, and the long-term stability is excellent. A typical total system accuracy is ±0.2°C.

A condensation hygrometer, unlike temperature or pressure sensors, must be in intimate contact with the gas to be analyzed. Thus, they are subjected to contamination. Often, a sampling system is recommended for applications involving considerable contamination (see Fig. 13.12). The set-up is also used to cool sample gas below 100°C. A contamination with a soluble salt, such as NaCl, leads to a systematic error. The dew-point indication will then be too high. This event can be readily explained, since a saturated brine solution exerts, at temperatures between 0 and 100°C, approximately 75% of the saturated water vapor pressure. It also means that dry salt gets wet when the relative humidity is higher than 75%. So, a mirror contaminated with NaCl exhibits a scattering of reflected light at a temperature higher than the true dew point.

General Eastern introduced a system to alleviate the effects of contamination. Intermittently, the mirror is cooled to a temperature well below the prevailing dew point. This causes an excess of condensing water. The salt is dissolved. On heating the mirror above the dew point, the water evaporates, leaving the redistributed salt in clusters. This effect is illustrated in Figure 13.13. Most of the surface is clean again. The effect is that the time between mirror cleanings becomes 10 to 100 times larger.

The frequency of the automatic intermediate cleanings is once per 2, 6, 12, or 24 hr. The time intervals between mirror cleaning could vary widely. In heavily contaminated areas, the mirror may have to be cleaned every day, whereas in clean environments the mirror could be operated without service for several weeks or months. In typical industrial applications with relatively clean air, the mirror is cleaned once or twice a month. Chilled-

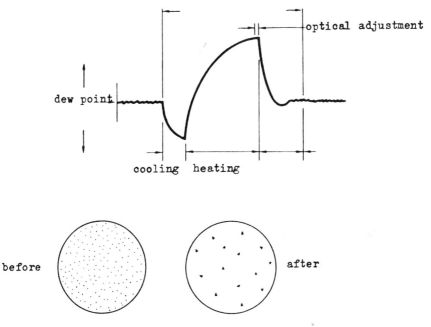

Figure 13.13 A PACER cycle for a condensation hygrometer. (Courtesy of General Eastern, Watertown, Massachusetts.)

mirror dew-point sensors are generally not suitable for operation in very corrosive environments. Special chilled-mirror sensors utilizing stainless steel housings and platinum mirrors can be supplied and are often used in mildly corrosive environments.

Condensation hygrometers are more expensive than saturated salt dew-point sensors and electrolytic hygrometers. However, they have superior properties and are, on the other hand, less expensive than spectroscopic instruments.

Ultraviolet (UV) spectroscopy

The principle of operation of a UV spectrometer is depicted in Figure 13.14. A xenon flash lamp is used to generate UV light. The radiation produced is transmitted down an optical fiber to the measurement path. The UV light then passes across the sample path to a reflector and is refocused onto a return fiber. The latter optical fiber conveys the radiation to a beam splitter where the light is divided into two channels—absorption and reference. Detection occurs by means of pmt detectors. The ratio of the two signals is proportional to the moisture in the intervening air sample. The background is that moisture in the air absorbs UV light at some wavelengths and does not do so at other wavelengths. How air is used to keep the optics clean, even in an adverse fouling environment, is depicted in Figure 13.15.

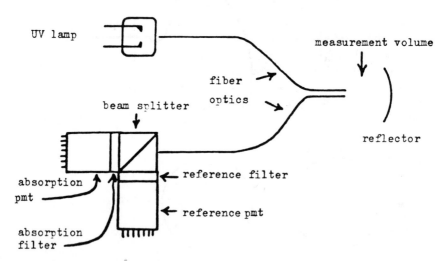

Figure 13.14 A schematic view of ultraviolet absorption with hygrometer optical components. (Courtesy of Pacer Systems, Inc., Billerica, Massachusetts.)

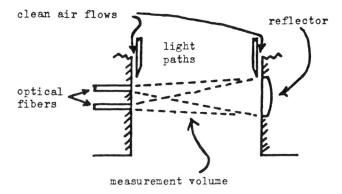

clean air flows

reflector

light
paths

optical
fibers

measurement volume

Figure 13.15 A schematic view of ultraviolet absorption with a hygrometer head. (Courtesy of Pacer Systems, Inc. Billerica, Massachusetts.)

In principle, infrared (IR) light can be used in much the same way as UV light. One aspect is that the measurement utilizing UV light remains accurate up to 500°C sample temperature. Background radiation in the IR band becomes a problem at high temperatures. It is sometimes held against UV spectroscopy that oxygen interferes. In principle, this is correct for the measurement (absorption) wavelength. However, the change in absorption that is due to a change in the oxygen concentration is under normal circumstances negligible in comparison to the absorption changes that are due to variations of the moisture content. Furthermore, the reference wavelength radiation is affected neither by oxygen nor by water.

UV spectroscopy is used in papermaking. Paper is dried indirectly on cylinder dryers. The evaporated water is removed by sweep air. The air is recycled in a loop and the hygrometer can be used to control the moisture content of the air in the loop. Corrosion is hardly an issue because a noncontact method is used. The parts coming into contact with the process fluid (probe) can be made out of stainless steel, e.g., grade AISI 316.

13.4 Dryer process control

Introduction

The production of dried material at minimal cost and maximum throughput is the objective of many drying operations. There is no need for automatic control as long as the mass balance and the process conditions do not change. However, changes do occur, and it is possible to compensate for their effect by means of process control, which comprises the measurement of a process variable and the subsequent automatic adjustment of a different process variable. The temperature of the exit gas is often measured and used as an input to a control loop. Often, the moisture content of the dried product can be inferred from this measurement. Furthermore, measuring temperatures is simple, accurate, reliable, and cheap. However, it is also customary to measure the moisture content of the dried material in an inferential way; e.g., a textile dryer can be controlled employing the conductivity of the exiting material as the input. Another example is the measurement of microwave absorption in food processing. Earlier in this chapter various on-line continuous moisture measurement methods were treated (see pp. 291–311). In principle, the analog signals of these instruments can be used for process control.

In many cases, the process variable acted upon is the heat supply or the feed to the dryer. Control of the airflow is hardly ever used. The background is that the air is sometimes important for the transport function (flash dryer) and that the air ducts are wide; hence, this would necessitate large and expensive control valves.

Many processes proceed in a hazardous area (electrical classification) and this is one reason to locate the control equipment in a control-room environment or "safe area." Consider the physical realization of a control loop measuring the exit-gas temperature and adjusting the natural gas flow to the burner (see Fig. 13.16): (a) measurement of the exit-gas temperature by means of a thermocouple, (b) converting the mV signal into a 4–20 mA signal, (c) processing the signal by means of an A/D-converter, which is part of a microprocessor (controller), and (d) sending an output signal 4–20

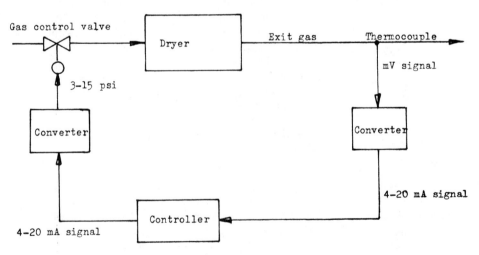

Figure 13.16   A typical control loop.

mA to a device converting the electronic signal into a pneumatic signal (3–15 psi) operating the control valve.

Control Systems

Two different control systems, feedforward and feedback control, can be distinguished.

Feedforward control

The principle is shown in Figure 13.17. Wet material continuously enters the dryer and when the flow or the moisture content changes, a different heating-medium flow is required. The controller takes care of adjustments. Feedforward control prevents an upset of the dryer from occurring; however, the actual result of the control action is not checked. Thus, small gradual shifts are often not noticed.

Feedback control

Figure 13.18 exhibits the principle. A temperature measurement serves to check the proper drying action. A change in this temperature initiates a controller action; for example, when the temperature decreases, more heat is sent to the dryer. This sounds simple; however, there are complications. By the time the temperature decrease is detected, some wetter material has already entered the dryer (delay time). Furthermore, at the moment the controller sends more steam to a heater, the mass of the heater starts heating up. It takes some time to heat the metal and hence to notice the

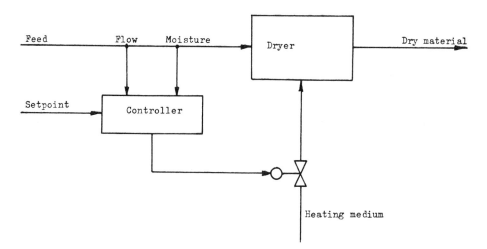

Figure 13.17   Feedforward dryer control.

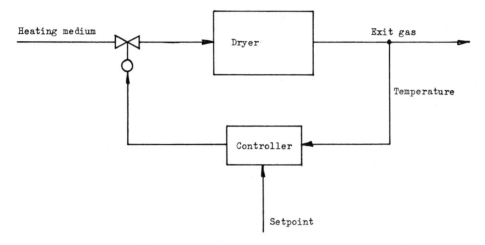

Figure 13.18   Feedback dryer control.

controller's effect of an increased drying-air temperature (resistance-capacitance lag).

One way to deal with the effects of delay time and resistance-capacitance lag is to tune the control loops under operating conditions. Ziegler and Nichols's rule is a practical means to attain this goal. A different avenue is the construction of a process model.

Advantages are sometimes found if feedforward and feedback controls are combined. Microprocessors and other process computers offer options to realize this. Three basic modes of continuous control exist: proportional action, integral action, and derivative action. The proportional action implies a corrective action that is proportional to the deviation of the measured value from the desired value (setpoint). Start from the situation that the desired value equals the setpoint. The actuator's position is defined. A deviation occurs (e.g., a temperature drop because of wetter feed), and the actuator (e.g., a valve) admits a larger heating-medium flow.

In the new situation, the desired value cannot equal the setpoint because then the original stem position would be resumed. The difference is called the *off-set*. The band width of the proportional action is an important variable. The integral action comprises a corrective action to undo the off-set. It sees to it that the desired value (actual value) becomes identical to the setpoint. The valve's stem gets a new position. The integral time is a variable.

The derivative action implies a control mode proportional to the rate of change of error. When a deviation increases rapidly, fast action is required and vice versa. The derivative time is the variable here. Derivative action enables anticipation. A three-term controller combines the three modes. In the past, three-term controllers often consisted of small boxes combined with a recorder. In conventional hard-wired control rooms, many of these PID-boxes can be found. Today's practice implies the use of microprocessors. A typical set-up is the combination of 8 control loops in one microprocessor.

The use of process computers does allow (if required) the use of more sophisticated (often tailor-made) algorithms than the P-, I-, and D- actions. Furthermore, the combination of control loops (as found in cascade control systems) is easier than with convential control equipment. The price of the components does not prevent the application.

A typical aspect of process computers is the process data-acquisition capability. Temperature, flow, and pressure data can be stored, displayed on visual display units (VDUs), plotted (a hard copy is obtained), and passed on to a supervisory process computer.

Control valves are almost exclusively activated pneumatically. A time lag of several seconds would prevail when the pneumatic lines are long. Electric signals are intermediate between the control room and the local pneumatic valve activation system.

Generally speaking, the process control of drying systems is usually not very sophisticated.

Continuous convective dryers

Three flows enter the drying system: the air, the heating medium, and the feed. The possibility to select feedforward control was discussed earlier. However, the majority of the continuous convective dryers have feedback control, which implies a check of the drying function and acting upon one of the flows entering the systems.

Flash dryer

The residence times of both the solids and the gas are extremely short, i.e., several sec. The hold-up of material is quite small. Rapid changes in the feed quantity or moisture content can upset the dryer. In practice, a situation often met is that the flash dryer is fed by a centrifuge or a filter. A centrifuged product's moisture content can gradually increase from 3 to 5 % by wt between wash-outs. The usual approach is to connect the exit-gas temperature to the heating medium flow (see Fig. 13.19). The exit-gas temperature signal can be used alternatively to alter the setpoint of a TC controlling the heater exit temperature. The latter set-up has a short response time.

Control by means of the feed is attractive because it implies a small response time. However, quite often the flash dryer must accept what the

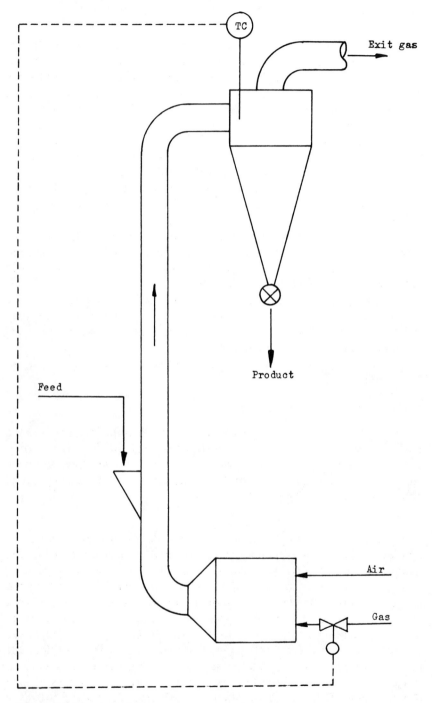

Figure 13.19 A flash dryer process control.

316

preceding equipment produces. Furthermore, controlling a flow of wet solids is not too easy, although it can, in principle, be done by means of a buffer and a rotary lock with variable speed.

Spray dryer

Here, too, the residence times are short. Connecting the exit-gas temperature to the fuel supply is quite common. The possibility of acting upon the feed flow in order to maintain the exit-gas temperature is also being used (see Fig. 13.20). However, this method cannot be used in combination with nozzle atomization since varying the feed flow would also influence the particle-size distribution.

A general remark pertaining to the flash dryer and to the spray dryer is that when the drying system is equipped with a steam heater, the resistance-capacitance lag which is due to the heater is considerable. Rapid changes of the feed quantity or moisture content can lead to "hunting" of the control system if the exit-air temperature acts on the steam flow. A connection of the temperature measurement to the feed flow may be necessary. Another option is the action of the exit-gas temperature on the steam pressure (higher pressure: higher condensation temperature).

Fluid-bed dryer

Both circular and rectangular dryers use relatively long residence times for the solids and short residence times for the gas. Rapid changes in the feed's quality or quantity will not readily upset the dryers because of the buffer capacity with respect to the solids. The usual control method for the rectangular fluid-bed dryer is to measure the product/gas temperature in the fluid bed just ahead of the cooling section and connect it to the heating medium supply (see Fig. 13.21). Now, the delay time is not very small. On the other hand, rectangular fluid-bed dryers experience axial dispersion leading to a relatively early temperature warning signal. Furthermore, as soon as hotter air is admitted (because the preceding centrifuge some time ago turned out wetter material), it acts upon the full hold-up immediately.

Usually, the dryer processes whatever the plant produces. Circular fluid-bed dryers are commonly controlled by connecting the air-exit temperature to the fuel supply. The time delay is nil here, and changes occur very gradually because of the bed's buffer capacity.

Rotary dryer

This dryer type can be controlled via the measurement of the exit-gas temperature which is then connected to the heating medium. A general remark for all continuous convective dryers equipped with a steam heater is that it is often more effective to increase the steam pressure than the

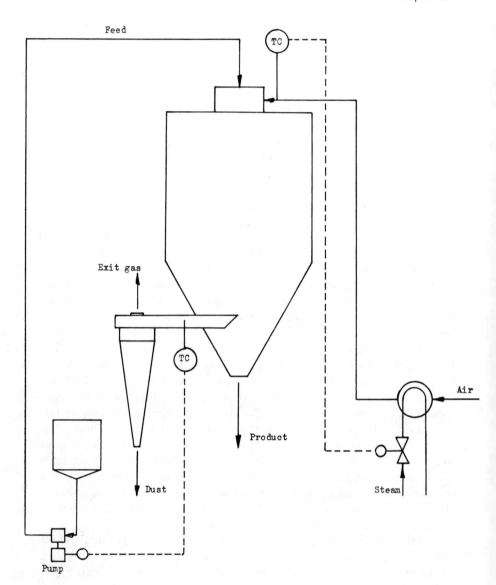

Figure 13.20   A spray dryer process control.

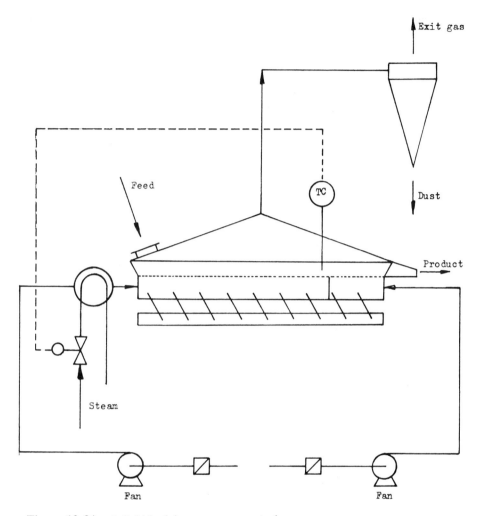

Figure 13.21   A fluid-bed dryer process control.

steam flow. A higher steam pressure gives a higher condensation temperature immediately.

Conveyor dryer

Often, conveyor dryers are equipped with a steam heater that heats the
make-up air indirectly. The temperature of the air circulating through the
product layer is measured, and this value is used to adjust the steam control
valve. The relative humidity of the recycled air is controlled manually by

means of a butterfly valve in the exit-air line. Here, the variation of the steam pressure can be quite effective.

Miscellaneous continuous convective dryers

Control of these dryers can be obtained by application of the methods that were discussed in this section. The Convex dryer can be compared to a flash dryer and the Rapid dryer to a rotary dryer.

The air flows through continuous convective dryers are usually adjusted manually (by butterfly valves). Generally, small underpressures in the drying chambers are preferred to avoid dust emissions.

Continuous contact dryers

The general approach is to measure the air-exhaust temperature and to use this signal to control the air-inlet temperature. The controller, e.g., can send more steam to the air heater or increase the steam pressure. The heating medium's temperature is usually kept constant by a loop controller. The setpoint can be adjusted, however, too high a medium temperature can lead to caking in some instances. The airflow per se (to remove the evaporated water) is adjusted manually.

Batch dryers

The control practice most commonly encountered is to regulate the heating medium temperature automatically and to measure the product temperature. The latter measurement can be used to terminate the drying step. A different possibility is to stop drying automatically when a certain amount of time has elapsed. The measurement of the product's temperature can be difficult. It is then often possible to infer the product's moisture content from other physical properties, such as, the vacuum. When the bulk of the moisture has been evaporated, the vacuum in a dryer decreases.

13.5 Energy recovery

Introduction

The 1980 world energy consumption amounted to 286×10^{12} MJ. 1 MJ (megajoule) corresponds with 10^6 Joule. The 1980 total was obtained almost exclusively from oil, gas, coal, uranium, and water. The contribution of other sources of primary energy like wind and sun radiation was negligible.

Baker and Reay (1982) give energy-usage data regarding drying in six industrial groups in England.

Subsector	10^9 MJ per annum		Percentage due to drying
	Drying	Total	
Food and agriculture	35	286	12
Chemicals	23	390	6
Textiles	7	128	5
Paper	45	137	33
Ceramic and building materials	14	127	11
Timber	4	35	11
	128 +	1103 +	12

The industrial groups were selected because drying is important in these sectors. Both primary and secondary energy sources were considered.

Strumillo and Lopez-Cacicedo (1987) stated that in United States industry, six dryer types account for about 99% of the total dryer energy consumption. In order of importance, these dryer types are (a) flash, (b) spray, (c) cylinder (indirect, papermaking), (d) direct-heat rotary, (e) indirect-heat rotary, and (f) fluid bed.

Heat can be recovered from two process flows; namely, the product and the exit gas. Heat recovery from the exhaust gas is met more often than heat recovery from the dried product. Hence the discussion will be restricted to the second type of efficiency improvement. However, two aspects in this field do not render things too easy:

1. The energy to be recovered is spread out in space (the exhaust gas is rather bulky).
2. The gases are not particularly hot.

The possible heat recovery methods are: (a) recycle of part of the exhaust air to the air inlet line, (b) indirect gas/gas heat exchange, (c) scrubbing, and (d) using a heat pump. It is possible to combine several methods. So, the advantages must be balanced against the investments required.

Sometimes the heat recovery is restricted to sensible heat recovery, however, latent heat can, in principle, also be recovered.

Exhaust-gas recycle

Part of the exhaust gas is recycled to the air-inlet line. There condensation may not occur. The recycle leads to a higher moisture level in the exhaust gas. Often, the capacity can be maintained only if the exhaust-gas temperature is increased. But such a measure counteracts the first step. Furthermore, for various reasons, it is necessary to rigorously remove dust particles from the recycle gas flow.

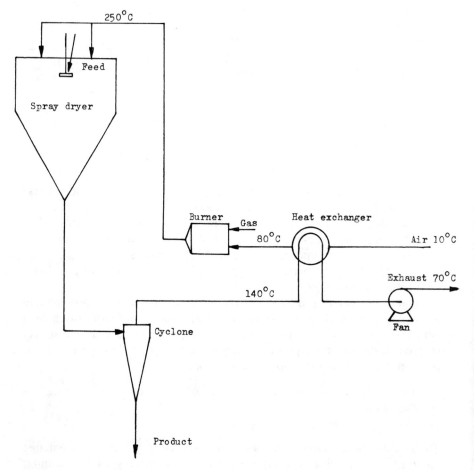

Figure 13.22   Heat recovery from a spray dryer using a gas/gas heat exchanger.

Indirect gas/gas-heat exchange

This procedure is depicted for a spray dryer in Figure 13.22. Usually, only sensible heat is removed. The heat exchanger can become contaminated on the exhaust-gas side. Consequently, the steps to counteract this are: (a) air filtering, (b) installing a glass heat exchanger (note: glass is fragile), or (c) periodic heat-exchanger cleaning.

Indirect gas/liquid heat exchange

Figure 13.23 exhibits a run-around coil installed to recover heat from the exhaust of a spray dryer. It can offer advantages over the former type of

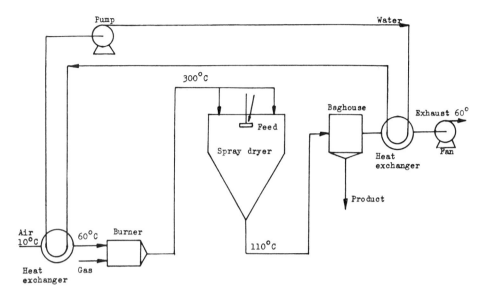

Figure 13.23 Heat recovery from a spray dryer by means of a run-around coil.

heat exchange because it is now not necessary to construct (so to speak) a fly-over crossing of roads (the heat exchanger).

Another possibility to recover heat concerns the installation of a heat pipe. The warm gas causes a thermal fluid to evaporate by indirect heat exchange. The vapor condenses in a different part of the conduit and transfers heat to the cold gas. It stands to reason that the distance between the two parts cannot be too large.

Scrubbing

The exhaust gas is scrubbed with cold water and warm water is obtained. This method can also recover part of the latent heat in the exhaust-gas flow. It is necessary to have an outlet for the warm water (e.g., process water). The thorough cleaning of the exhaust gas is a bonus of this method.

Heat pump

A heat pump can accept thermal energy at a low level (both sensible and latent heat) and deliver the energy at a high level. This is accomplished by means of mechanical energy. A thermal fluid acts as an intermediate. The thermal fluids in use at present cannot get warmer than approximately 100°C because of their thermal stability.

References

Baker, C. G. J. and D. Reay (1982). Energy usage for drying in selected U.K. industrial sectors, *Transactions 3rd International Symposium on Drying,* Marcel Dekker, New York.

Klein, A. (1987). Comparison of rapid moisture meters, *Min. Proc.,* 28, 10.

Strumillo, C. and C. Lopez-Cacicedo (1987). Energy aspects in drying, *Handbook of Industrial Drying* (A. S. Mujumdar, ed.), Marcel Dekker, New York.

14

GAS/SOLID SEPARATION METHODS

14.1 Introduction

The movement of gases is inherent to drying operations. The gases may entrain particulate material and a gas/solid separation step is usually required. Convective drying implies intimate contact between the air used for drying and the material being dried; this is not the case for contact drying. Some dryers have a preseparation of the gas and the solids in the dryer itself, e.g., fluid-bed and rotary dryers. Preseparation does not occur in a flash dryer, whereas spray dryers can operate in both modes. In conveyor dryers, the solid remains stationary and usually only low levels of entrainment are found. Cyclones are capable of operating with high inlet dust levels (e.g., 200 g/m^3) and it is common to install them upstream of such other collectors as bag filters or scrubbers. However, they are not suitable for the efficient collection of fine dusts.

Because of their positive nature, fabric filters will generally ensure a high collection efficiency (greater than 99%) even on submicron-sized dusts. Their main limitation is the heat sensitivity of the filter media. Natural fibers cannot be used at temperatures exceeding approximately 80°C whereas synthetic fibers can be used up to 150°C. Glass fiber will withstand continuous operation up to 260°C. A low-energy scrubber can collect fine soluble particles effectively and fine insoluble particles can be collected using high-energy scrubbers; however, the energy requirement increases exponentially in the submicron range as the particle size becomes smaller.

Cyclones, fabric filters, and scrubbers are widely used in convective drying plants and are available in packaged assemblies.

Contact dryers are frequently equipped with a small fabric filter, which is suitable to retain the small particles that are entrained with the low gas velocities in contact dryers. It is important to avoid condensation in dust collectors since filter fabrics are easily blinded and are then difficult to clean. Cyclones can also become plugged.

14.2 Cyclones

Introduction

A cyclone is a device that accomplishes a solid/gas separation by imparting a spinning motion to the two-phase system. The particles move outward to the wall of the cyclone because of the centripetal force and subsequently travel downward to enter a discharge hopper. The gas that is freed from solids leaves at the top through a central outlet. Figure 14.1 shows a typical cyclone where the dust-containing gas enters the cyclone body tangentially and moves downward in a vortex motion. The flow direction reverses at the apex of the cone, and the cleaned gas travels upward in a second vortex.

Sometimes it is stated that a cyclone is a cheap centrifuge and certainly an element of truth is contained in this statement. However, it must be recognized that cyclones generate much lower radial acceleration than centrifuges, e.g., a cyclone with a diameter of 1 m and an inlet velocity of 20 m/sec has a radial acceleration (expressed as a multiple of the acceleration due to gravity) of $20^2/(0.5 \times 9.81) = 81.5$. This acceleration is considerable but is not spectacular, since peeler centrifuges typically operate with an acceleration figure in the range 500–1,000. A cyclone with a diameter of 2 m and the same inlet velocity as above has a radial acceleration of 40.75. Consequently, large cyclones are less efficient than small ones. However, the pressure drop is, in principle, equal for cyclones if their geometric ratios are equal. These factors provide an incentive to arrange several cyclones in parallel operation rather than use only one large one.

Cyclones are cheap devices for dust collection but their main limitation is their poor efficiency in collecting particles smaller than about 5 microns. High dust loadings can be handled and hence cyclones can be used to separate flash-dried particles from the exiting gas.

Sizing and process data

Koch and Licht (1977) summarized proportions investigated in a number of studies. Cyclones typically have diameters in the range of 0.1 to 2 m (The Institution of Chemical Engineers, 1985). Cyclones can be sized using the rule of thumb that the normal flow rate = $1.5 \times D_c^2$ m³/sec (The Institution

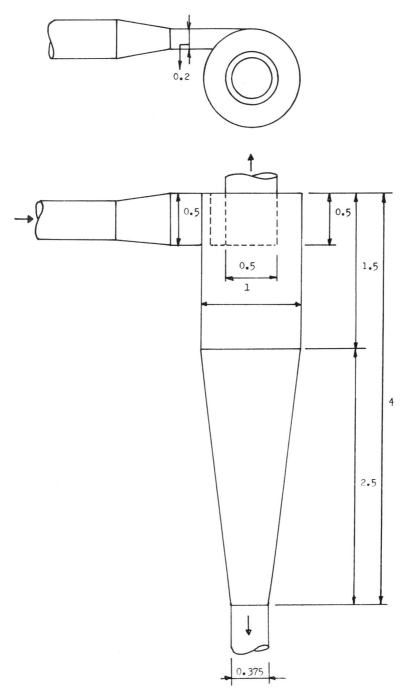

Figure 14.1 Cyclone for gas/solids separations. Figures indicate proportions.

of Chemical Engineers, 1985) for a high efficiency, medium throughput pattern. Inlet velocities are normally in the range 10–25 m/sec and, for a given cyclone, the efficiency increases when the inlet velocity increases. However, entrance velocities in excess of 30 m/sec cause turbulence which, in turn, leads to bypassing and re-entrainment of separated particles and hence a decrease in efficiency.

Pressure drops of between 50 and 150 mm in a water gauge are normal. Koch and Licht (1977) recommended Shepherd and Lapple's approach (op. cit.) to estimate the pressure drop in a cyclone. First, the number of velocity heads is calculated

$$N_H = K(a \times b/D_e^2) \tag{14.1}$$

$K = 7.5$ if a neutral inlet vane is present and 16 if an inlet vane is absent. Second, the pressure drop is calculated:

$$\Delta P = 0.0512 \times \rho_g \times V_i^2 \times N_H \quad \text{mm wg} \tag{14.2}$$

Worked example

Geometrical ratios cyclone

$a = 0.5 \times D_c$
$b = 0.2 \times D_c$
$D_e = 0.5 \times D_c$

$a \times b/D_e^2 = 0.5 \times 0.2/0.25 = 0.4$
$K \times 0.4 = 16 \times 0.4 = 6.4$ (no inlet vane)
Take $\rho_g = 1.0$ kg/m^3 and $V_i = 20$ m/sec
$\Delta P = 0.0512 \times 1.0 \times 400 \times 6.4 = 131$ mm wg

Koch and Licht (1977) recommended relationships for the calculation of collection efficiency. A potential source for poor cyclone performance is the air ingress into the dust outlet of the cyclone.

14.3   Fabric Filters

Introduction

Fabric filters are devices that perform gas/solid separation by means of a fabric, with the gas phase passing through a cloth and the particulate matter being retained. The structural enclosure containing the bag-shaped filter is termed a *baghouse* (see Fig. 14.2, which schematically shows a typical baghouse). The dust-ladened air enters via an inlet pipe and flows into the bags, where the dust accumulates inside the bags and the gas passes through the dust layer and the cloth and leaves the filter through the outlet

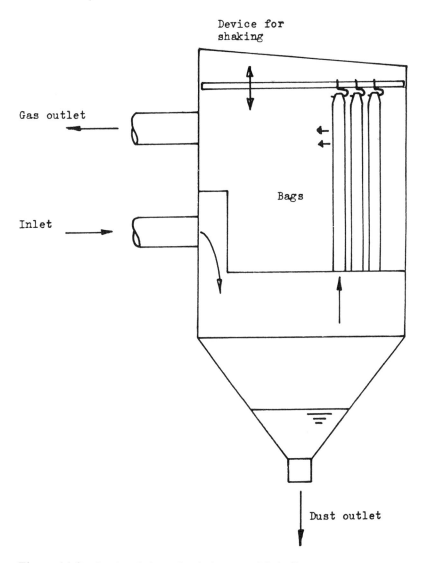

Figure 14.2 Sectional view of a shaker-type fabric filter.

pipe. Intermittently, the bags are shaken and the solids fall into the hopper. Fabric filters are very effective and even on submicron-sized dusts collection efficiencies of greater than 99% can be obtained.

The thermal stability of available fabrics imposes the greatest limitation on filter applications. Natural fibers (cotton and wool) cannot stand long-term exposure to temperatures exceeding 80–90°C, whereas synthetic and glass fibers can be used at higher temperatures; e.g., polyester can withstand temperatures up to 140°C. Because of this temperature limitation, fabric filters are mainly used in drying systems for organic chemicals, where drying air-inlet temperatures generally do not exceed 200°C and the exit gases are usually not above 100°C. However, the possibility of noncooled drying air passing directly into the baghouse must be evaluated. Baghouses must be safeguarded against dust explosions because dust is present and considerable dust hold-up in the baghouse can occur. Thorough bonding is necessary, and metal- or carbon-coated fibers are often included into the fabrics. Explosion panels that relieve directly or through short lines to the atmosphere are normal features of baghouses.

Fabric filters are very flexible and are able to handle wide variations in gas flow, dust concentration, and particle size. The superficial face velocity is an important independent process variable that is defined as the velocity in m/sec at which the gas flows through the fabric. The User Guide (The Institution of Chemical Engineers, 1985) provides a table containing recommended average filtration velocities for many dusts. Generally, the separation of a fine dust is carried out at a lower velocity than a coarse dust. Values between 0.01 and 0.03 m/sec are applicable for woven media, whereas the range is 0.02 to 0.06 m/sec for felted media.

Pressure loss is an important dependent process variable. In many instances, baghouses are split up into sections in order that individual sections can be taken off-line sequentially for cleaning while the other compartments continue in operation. Hence, the filter pressure fluctuates but does not exceed 150-mm water gauge. Normally, the pressure loss often automatically triggers the cleaning of a section. The pressure drop across the filter can be expressed as

$$\Delta P = \Delta P_1 + \Delta P_2 \tag{14.3}$$

ΔP_1 is the pressure drop across the fabric itself after having been in use for some time. ΔP_2 is the pressure drop that is due to flow through the dust deposit. ΔP_2 increases with time. Experience has taught that the above equation can also be expressed as

$$\Delta P = C_1 \times v + C_2 \times v \times m \tag{14.4}$$

m is the mass of the dust deposited per unit area (m is a function of time). C_1 and C_2 must be obtained experimentally.

Fabrics

Fabrics are either felted or woven. The traditional method for making a fabric is weaving, in which a series of yarns are interlaced at right angles to

Figure 14.3 A plain weave for fabric construction.

produce a plain, twill, or satin weave (see Fig. 14.3). Weaving is followed by other treatment, e.g., surface finish. Felted fabrics consist of fibers that are compressed under high pressure. The gas velocity through felted fabrics is higher than that through woven fabrics; hence, baghouses equipped with felted media are smaller than those provided with woven fabric. Furthermore, felted fabrics are more efficient than weaves, and these two factors have led to an increased usage of felted fabrics. However, felted fabrics require a more rigorous cleaning method than woven fabrics.

Bags are typically 0.1 to 0.3 m in diameter and 2 to 10 m in length. A section of a multicompartment filter provides an area of 10 to 20 m^2.

Initially, the cloth is the filter medium; however, the surface layer generally becomes the dominating filter medium once it has formed. Kraus (1979) reported an example of two dissimilar fabrics giving the same performance because of this phenomenon. The resistance against acids and alkalis should also be checked. Perry et al. (1984) provided much information concerning the properties (including mechanical) of fibers for fabrics. Glass can be used at a relatively high temperature; however, the finish limits the maximum temperature range. Glass has poor flex-abrasion qualities and hence cannot be used in filters that are cleaned by means of shaking.

It is important that the exit gas is maintained above its dew point otherwise condensation within the baghouse can spoil the fabric. Table 14.1 provides data on the thermal sensitivity of the filter media.

Table 14.1 **Maximum Recommended Operating Temperatures for Common Fabrics**

Fabric	Operating temperature (°C)
Cotton	80
Wool	90
Nylon	90
Polyacrylonitrile	125
Polypropylene	90
Polyester	140
Teflon	230
Glass	260

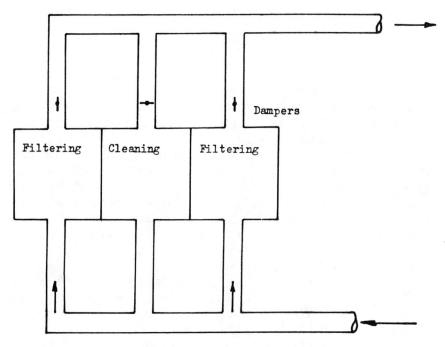

Figure 14.4   A three-compartment bag filter: one compartment is being cleaned.

Cleaning methods

It is customary that filters are classified according to the cleaning method. The three methods that are used are shaking, reverse airflow and reverse air pulse.

Figure 14.2 shows a filter that is cleaned by shaking, which was the earliest method to be developed. The flow in the bags is outward and the bags are commonly made of woven fabrics; felted media require a more powerful cleaning method, e.g., reverse air pulse. Large filters consist of sections that are cleaned sequentially off-line as shown in Figure 14.4. Filters equipped with reverse air flow often have bags made from woven glass since reverse airflow provides a fairly gentle cleansing action. The section to be cleaned is isolated from the process by means of a damper, and clean air passes through the fabric in the reverse direction. Here too, the dust accumulates inside the bags.

Baghouses that are cleaned by reverse air pulse operate with an inward gas flow and are usually equipped with felted fabric bags. Often, the filters are not split up into sections and cleaning (see Table 14.2) is performed on-line. The air pressure for pulse cleaning is between 4 and 7 bar. Figure 14.5 illustrates the principle of this cleaning process. Figure 14.6 shows an outdoor version of this type of filter as part of a Schugi drying installation.

14.4 Scrubbers

Introduction

A scrubber is a device that uses a liquid to clean a gas. The material that is removed from the gas can be either in a liquid phase (droplets) or in a solid phase (particles). In drying plants, scrubbers remove solids and are often preceded by cyclones.

A good contact between the liquid and the gaseous phase must be obtained; to this end, the scrubbing liquid is dispersed in a spray or spread in a film over the internal surfaces of the scrubber. Usually, the contactor is followed by an entrainment separator. The two most important basic collection mechanisms in all scrubbers are centrifugal deposition and inertial impaction and interception.

The centrifugal deposition mechanism comprises spinning out of particles by centripetal force caused by a change in gas flow direction. This is the mechanism prevailing in cyclone separators and is effective on particles down to approximately 5 microns.

Inertial impaction occurs when a gas stream flows around a small object and the particles suspended in the gas stream continue to move toward the

Table 14.2 Survey of Cleaning Methods for Fabric Filters

Cleaning method	Fabric type	Flow direction[a]	Dust accumulation[b]	Cleaning	Single-/multi-compartment	Efficiency[c]	Gas velocity m/sec
Shaking	Woven	Outward	Inside	Off-line	m	+	0.01–0.03
Reverse airflow	Woven (glass)	Outward	Inside	Off-line	m	+	0.01–0.03
Reverse air pulse	Felted	Inward	Outside	On-line	s	++	0.02–0.06

[a]Flow direction: *outward* means flow from the inside of the bag to the outside.
[b]Dust accumulation: *inside* means dust accumulation in the bag.
[c]The efficiency concerns the filtration efficiency and not the cleaning efficiency.

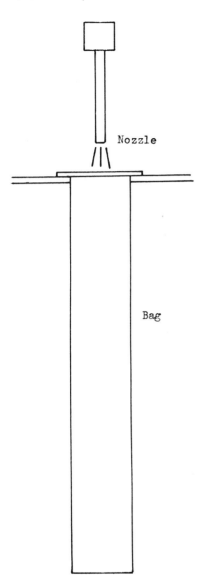

Figure 14.5   Cleaning by means of a reverse air pulse

Figure 14.6   A reverse air-pulse filter. (Courtesy of Schugi Process Engineers, Lelystad, The Netherlands.)

object and eventually are trapped. This mechanism is effective for particles as small as 0.5 microns.

Actually, there is no fundamental difference between these two collection mechanisms. The first one is concerned with large-scale flow direction changes, whereas the second concerns small-scale changes.

Brownian diffusion becomes important for particles that are smaller than 0.3 microns. This mechanism becomes quite effective for very small particles, and the performance depends upon the residence time. The first two collection methods are only slightly dependent on the residence time. Hence, multiple-stage contacting has no real advantage over single-stage contacting for the first two collection methods. Semrau (1977) described that, for a given particulate, there is a fundamental relationship between the collection efficiency of a scrubber and the energy dissipated in the gas-liquid contacting process. It appears that the efficiency is relatively independent of scrubber geometry and also hardly depends on the way the power is applied.

A distinction is often made between low-, medium-, and high-energy scrubbers. High-energy ones have pressure drops exceeding 500 mm water gauge and can collect particles as small as 0.5 microns; however, their running costs are high. Low-energy types have pressure drops of up to 100 mm water gauge.

It is not clear whether or not wettable particles are more readily collected than nonwettable particles or whether the use of wetting agents promotes collection. Normal liquid to gas ratios are between 0.1 and 10 m^3 of liquid per 1,000 m^3 of gas. Solid concentrations in the recycle scrubbing liquid can be as high as 10% by weight. Calvert (1977) indicated that the scrubber types most frequently used for drying systems are centrifugal, packed bed, moving bed, and plate. These scrubbers are all low- or medium-energy systems. Systems into which the liquid is sprayed are also used. These five scrubber types will now be considered in some detail.

Centrifugal scrubbers

A centrifugal scrubber is a low-energy system. Figure 14.7 shows a typical "wet cyclone." In the absence of a spray, a particle cut diameter of approximately 5 microns can be obtained.

The gas being cleaned is given a spinning motion by means of a tangential inlet port or by providing stationary swirl vanes in the collector. Normally, cyclone inlet velocities are in the range of 15–25 m/sec, but inlet velocities as high as 30 m/sec can be used in wet cyclones.

Centrifugal scrubbers are relatively simple and do not have small passages; hence there is little risk of plugging. Entrainment separation is taken care of within the scrubber itself.

Packed-bed scrubbers

Figure 14.8 schematically shows a tower scrubber. Normally, packed-bed scrubbers are used for the removal of soluble particles since high insoluble dust loadings can cause plugging. The scrubbers may be packed with, for example, Raschig rings or Berl saddles and gas/liquid contacting is performed in a countercurrent manner.

A 1-in packing size corresponds with a cut diameter of approximately 1.5 microns. Typical superficial gas velocities are of the order of 2 m/sec and droplet entrainment is prevented by means of a demister.

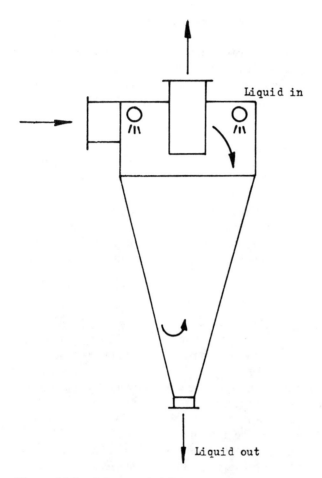

Figure 14.7   A "wet cyclone"

Moving-bed scrubbers

A moving-bed scrubber is depicted schematically in Figure 14.9. Gas passes upward through the suspension of plastic or glass spheres in the cleaning liquid and the continuous movement prevents the formation of deposits. Thus, moving-bed scrubbers offer advantages when there are known problems with scale deposits. Very efficient moving-bed scrubbers will retain particles as small as 1 micron. The pressure loss is approximately equal to the hydrostatic pressure that is due to the liquid bed height. The liquid is added from either the top or bottom of the bed. An entrainment separator prevents liquid extraction by the leaving gas.

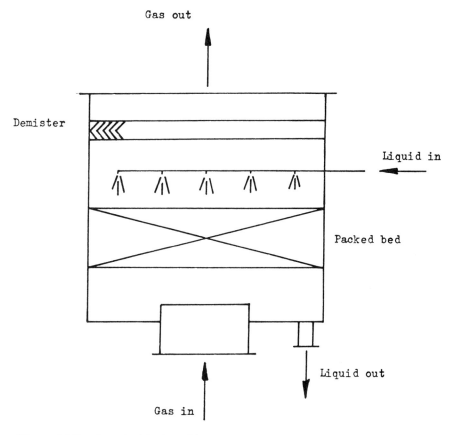

Figure 14.8 A packed-bed scrubber.

Plate scrubbers

A plate scrubber is shown schematically in Figure 14.10. Pressure drops are in the range 150–400 mm water gauge and a typical superficial gas velocity is 3 m/sec. Particles down to approximately 1 micron can be collected efficiently. The collection efficiency of sieve plate scrubbers increases as the perforation diameter decreases. Systems that have a tendency toward scaling require special plate design to avoid solids adhering to the plate.

Spray scrubbers

Two types of spray scrubber exist: in one the liquid is sprayed into the gas which is being cleaned and in the second type the motion of the gas stream

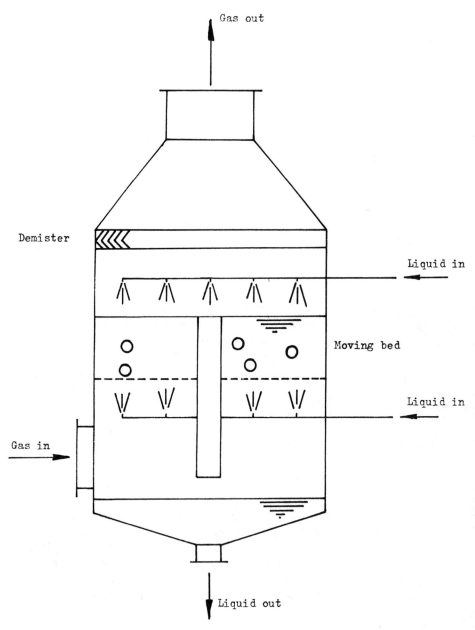

Figure 14.9   A moving-bed scrubber.

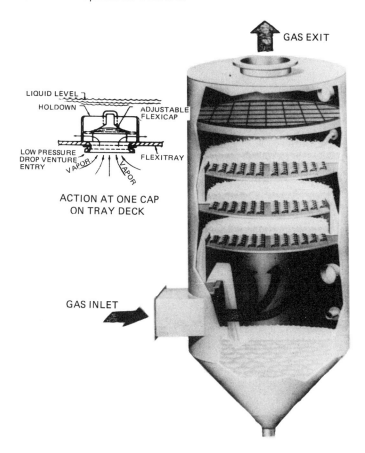

GAS EXIT

LIQUID LEVEL
HOLDOWN   ADJUSTABLE
                FLEXICAP

LOW PRESSURE
DROP VENTURE   FLEXITRAY
ENTRY   VAPOR   VAPOR

ACTION AT ONE CAP
ON TRAY DECK

GAS INLET

Figure 14.10   A plate scrubber. (Courtesy of Koch Engineering Company, New York.)

is utilized to atomize the liquid into droplets. The latter type employs high gas velocities in order to raise the relative velocity between the gas and the liquid.

Notation

a	Cyclone inlet height	m
b	Cyclone inlet width	m
C_1	Cloth/dust system constant	$N \cdot sec/m^3$
C_2	Dust deposit system constant	1/sec
D_c	Cyclone body diameter	m

D_e	Cyclone gas outlet diameter	m
K	Cyclone pressure-loss constant	
m	Mass of dust deposited per unit filter area	kg/m^2
N_H	Number of velocity heads	
v	Filtration superficial face velocity	m/sec
V_i	Cyclone inlet velocity	m/sec
ΔP	Pressure loss	N/m^2 or mm wg
ΔP_1	Pressure loss across the "used" fabric	N/m^2
ΔP_2	Pressure loss across the dust deposit	N/m^2
ρ_g	Gas specific mass	kg/m^3

References

Calvert, S. (1977). How to choose a particulate scrubber, *Chem. Eng.,* 84, 54.

Koch, W. H. and W. Licht (1977). New design approach boosts cyclone efficiency, *Chem. Eng.,* 84, 80.

Kraus, M. N. (1979). Baghouses: separating and collecting industrial dusts, *Chem. Eng.,* 86, 94.

Perry, R. H., D. W. Green, and J. O. Maloney (1984). *Perry's Chemical Engineers' Handbook,* McGraw-Hill, New York.

Semrau, K. T. (1977). Practical process design of particulate scrubbers, *Chem. Eng.* 84, 87.

The Institution of Chemical Engineers (1985). *A User Guide to Dust and Fume Control,* The Institution of Chemical Engineers, Rugby, England.

15

DRYER FEEDING EQUIPMENT

15.1 Introduction

Feeders are defined as various devices that introduce a variety of materials directly into dryers. Typically, the feeder is located at the interface between materials-handling equipment and the dryer. Some feeds can be pumped; for example, all spray-dryer feeds are pumped through a nozzle or a rotary atomizer. Many wet cakes can be made to flow and can thus be pumped to a vertical thin-film evaporator or a drum dryer. But in the majority of cases, the feed cannot be pumped and specialized equipment for the handling of bulk solids is used for storing, transporting, dosing, and controlling dryer feeds.

Often large storage containers are required because the feed is transported over long distances and is then stored in a bin or hopper. The design of the storage container must be based upon the flow properties of the material being handled. If the dryer feed is received directly from equipment that operates batchwise, then an intermediate storage container is required. These storages are deliberately not oversized in order to reduce the likelihood of setting and caking. However, if the dryer feed is received continuously from a processing plant, then a buffer, in principle, is not required even though process control factors may necessitate its use. Dryers with small product hold-up, for example, flash dryers and vigorously agitated, horizontal-contact dryers can be seriously affected by feed variation and interruption.

Intermediate storages are often combined with feeding or dosing equipment, such as a bottom-mounted screw in a hopper. Transport devices in

drying systems are identical to those that are used for the transport of bulk
solids; but a selection must be made because not all transport equipment is
suitable for handling wet solids. The most commonly used systems are
screw, belt, and vibrating conveyors. Each of these systems can be utilized
for dosing duties, but belt weighers and loss-in-weight feeders are the most
reliable and accurate systems. The weight is measured by means of strain
gauges, and the signal can be used for process control by means of a
feedback control loop. Thus, variations in bulk density of the feed can be
compensated.

A loss-in-weight feeder can be a bin with a bottom-mounted twin screw.
The bin's weight is measured by use of strain gauges, and the signal is used
to adjust the rotational speed of the twin screw. The steady flow of material
to the screws is promoted by a slowly rotating agitator.

Screws, rotary locks, and vibrating conveyors are also used for dosing;
the rotational speed (of screws and rotary locks) or frequency (of vibrating
conveyors) are related to the throughput. However, if the bulk density
varies, a recalibration of the system should be undertaken which is often
impractical because of the frequency with which changes occur. Thus, an
absolute feed rate is frequently not known. Nevertheless, such systems can
be used; for example, the exit-gas temperature in a flash dryer can be used
to regulate the rotational speed of a rotary lock so that a steady exit
temperature is maintained.

Often, it is important to prevent gas from flowing through feeders since
ingress of hot gases from the dryer may give rise to incrustations and thus
lead to losses, whereas cold air flowing through the feeder to the dryer
might interfere with the control system if the gas temperature is used for
process control. Another hindrance that could spoil the dryer operation is
the ingress of air or condensation in the gas vent system. If a screw feeder is
being considered, it is usually advantageous if one to one and a half pitches
of the screw flight be left off at the discharge end of the feed screw. The
choke formed by material moving poorly will often minimize air leakage
(Komline-Sanderson Engineering Corporation, 1988).

Many feeders become hot during operation, which is sometimes acceptable, but in other instances it may be necessary to provide indirect cooling.

Sections 15.2 through 15.11 deal with feeders to various dryer types,
with the important convective dryers covered first.

15.2 Fluid-bed dryers

In circular fluid-bed dryers, distribution and acceptance of the feed is relatively easy because the bed is deep and well mixed. But it is important that
the feed not be introduced near a wall, i.e., a screw must protrude into the

freeboard over the bed and then release the feed. Normally, the feeder is not subject to heating because the gas that flows past it has been cooled by the bed. However, contact heating may create an effect, even though the distance between the hot air plenum and the feeder is normally several meters.

Gas leakage can be either into or out of the dryer. Leakage into the dryer does not influence the thermal efficiency because the vented gas is not used for drying purposes; however, it can depress the gas-vent temperature and if this is used for process control, the process can be disturbed. Leakage out of the dryer generates a dust problem and may introduce incrustations in the feeder.

Usually, screw and rotary lock feeders are used in circular fluid-bed dryers. In rectangular fluid-bed dryers, if the bed is relatively deep, e.g., 0.5 m, distribution and acceptance of the feed is comparatively easy. However, if the dryer has a shallow bed, the feed must be distributed over the full width of the dryer. In these circumstances, it is customary to receive the output from the feeder on a chute that has triangular baffles (like a sample splitter). Vibrating fluid-bed dryers have shallow beds so that the amount of material being vibrated is minimized; also the vibrating action prevents incrustation in the air distributor. However, sometimes special measures to facilitate feed acceptance are necessary: more air needs to be passed through the feed section or a relatively low temperature must be used for the drying gas to prevent incrustations. The information given above on feeder temperature and gas leakages for circular fluid-bed dryers is equally applicable to rectangular fluid-bed dryers. Some examples of feeder systems that are used with fluid-bed dryers are as follows:

1. Sodium chlorate crystals containing 3% water by wt and coming from pusher centrifuges are fed to a circular fluid-bed dryer by means of a screw (Vreeland and Bacchetti, 1983). The gas-vent temperature is kept at 250°F (121°C) by adjusting the drying-gas temperature within the range 400–450°F (204–232°C).

2. Vacuum pan salt containing 3% water by wt is fed to a circular fluid-bed dryer by means of rotary locks. The gas-vent temperature is kept at 235°F (113°C) by adjusting a valve in the gasline to the furnace (Imes and Jobes, 1963).

3. Vacuum pan salt containing 3% moisture by wt and coming from pusher centrifuges is transported to a rectangular fluid-bed dryer by means of conveyor belts. Feeding occurs by means of a vibrating feeder, the feed entering the dryer via a chute with triangular baffles. Process control is achieved by keeping the highest bed temperature constant. There is a small underpressure in the freeboard, and air leaks

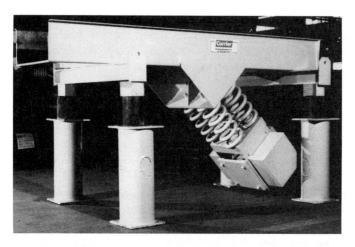

Figure 15.1   A vibrating feeder. (Courtesy of Carrier Vibrating Equipment, Inc., Louisville, Kentucky.)

into the dryer at the feeding point. Figure 15.1 depicts a vibrating feeder, driven by a 1-hp motor, employing the natural-frequency principle. It has a width of 2 ft and a length of 7 ft 6 in.

15.3   Direct-heat rotary dryer

Three feeders are frequently encountered, namely, chute, screw, and vibratory. If the dryer operates countercurrently, the heat load of the feeder is limited. Although ingress of air does not reduce the feeder's economy it does depress the vented air temperature, and this can interfere with the operation of the control system. Air leaking out at the feeding point creates a dust problem. If the dryer operates cocurrently, the feeder may become overheated and a special construction or indirect cooling with water may be required. Ingress air reduces the heat economy and the reverse gives rise to dust problems.

15.4   Flash dryer

Flash dryers are used for free-flowing powders, granular and crystalline solids, slurries, and pastes. The latter two feeds can be accepted only if the product is recycled to make the feed acceptable for handling. The maximum particle size should not exceed 1–2 mm because the particles must be transported vertically and larger particles often require longer residence times than can be afforded by flash dryers.

The distance between the feed stream and the wall is small at the point at which the feed enters and the entering feed is not mixed with material already present in the feed section. These two aspects distinguish flash dryers from the other types of convective dryers and call for careful consideration of the properties of the feed entering the dryer.

The concept of a flash dryer looks simple since drying and vertical transportation are combined. However, air takes care of the vertical transport in a direction opposed to gravity, because there are no mechanical means for transport. Often, special devices must be integrated into the flash-dryer system to ensure proper feed acceptance and processing. The main types of equipment are mixers, mills, slings, and classifiers. A mixer can mix feed and dry product to improve the flow characteristics of the feed (see Fig. 7.1 in Chapter 7). Figure 15.2 depicts a double-paddle mixer suitable for this procedure, which is often termed *backmixing.* A mill disintegrates the feed to enable good dispersion into the gas stream. The mill can physically take the form of a sling, a relatively simple mill, or a sophisticated mill (refer again to Fig. 7.1, which shows a cage mill). A sling is often installed at the lowest point of the drying system, since any coarse material is reduced in size and can then be entrained by the air. The sling can be installed if the feed is introduced using a screw or a rotary lock.

A classifier is located at the top of the dryer tube where it will separate coarse material that is subsequently recycled to the feed via an in-line mounted mill. So, although a flash dryer looks simple, a set of auxiliary devices is often required to guarantee solid handling. The drying of filter cakes, for example, usually requires a mixer and a cage mill. The operation of a flash dryer is cocurrent and this implies that the feeder (cage mill, sling, screw, or rotary lock) can become overheated. Hence, special solutions must sometimes be applied such as an indirect shaft cooling of a mill.

Figure 15.2 A double-paddle mixer. (Courtesy of C-E Raymond, Combustion Engineering, Inc., Chicago, Illinois.)

Ingress air at the feed point adversely affects the heat economy whereas hot gas leaking out of the dryer may create a dust problem and lead to incrustations in the feed equipment.

Examples of the use of feed systems in flash dryers are as follows:

1. Anhydrous sodium sulfate crystals containing approximately 5% water by wt are fed to a flash dryer by means of a screw where the feed is not mixed with previously dried material. A sling is installed at the lowest point of the drying system to disintegrate larger material coming down. Coarse material is recycled from the dryer head into the feed screws via an in-line mounted mill.
2. A filter cake containing approximately 25% moisture by weight is mixed with previously dried material and is fed by means of a screw to a sling at the lowest point. Recycle of milled coarse material is carried out.

15.5   Spray dryer

Generally, the design practices used for the storage and transport of solutions and slurries are applied. It should be noted that on introducing the feed, the distance between the feed and the walls is rather large and, in this respect, spray dryers differ from the other important convective dryers. The following examples are feeder systems that are used with spray dryers:

1. Two process vessels are used to prepare and store a slurry which then passes through a colloid mill and is then pumped through to a rotary atomizer. The colloid mill protects the atomizer against blockages.
2. A series of two pumps transports a solution to a single-fluid nozzle. The first pump is a low pressure one whereas the second pump generates the pressure required for atomization.

15.6   Conveyor dryer

Two principal classes exist in conveyor dryers: one in which the material to be dried is in the form of rather large particles (e.g., bread crumbs, nuts, and tobacco). In the second class, the material to be dried must be physically formed. This procedure is often carried out in the chemical industry when filter cakes are processed using conveyor dryers.

Various types of equipment can be used in the first case, e.g., the oscillating spreader (Fig. 15.3) and the wiper feed (Fig. 15.4), both of which spread the material to be dried evenly across the full width of the moving conveyor at the feed end.

Figure 15.3 An oscillating spreader. (Courtesy of Proctor & Schwartz, Inc., Horsham, Pennsylvania.)

A simple feeder is a hopper, but its application is limited to hard spherical objects that flow easily, such as nuts (Proctor and Schwartz, 1988).

Special feeders have been developed for tobacco and cereals. Vibrating feeders can be used with granular materials. Several techniques can be used for feed preforming. Perry et al. (1984) gives a review of the methods that are employed: (a) preforming by means of steam-heated finned drums, (b) preforming of thixotropic filter cakes by scoring with knives, and (c) briquetting and extrusion.

Rolling extruders are quite commonly used in the chemical industry to preform filter cakes. The material must remain sufficiently fluid after dewatering for this technique to be successful. Figure 15.5 illustrates a rolling extruder. The width of the trough corresponds to the width of the conveyor dryer and the filter (e.g., a rotary vacuum filter). The arms containing the rolls reciprocate and the rolls roll over the perforated trough wall while the paste is pushed through the holes forming extrudates. If the extrusion pressure is too low, the extrudates are mechanically weak which can lead to dust formation. An extrusion pressure that is too high results in extrudates having poor dispersion properties or even dewatering in the forming equipment, which may cause jamming. The pressure is influenced by the water content of the cake, the cake properties, and the hole sizes along with the percentage of free area.

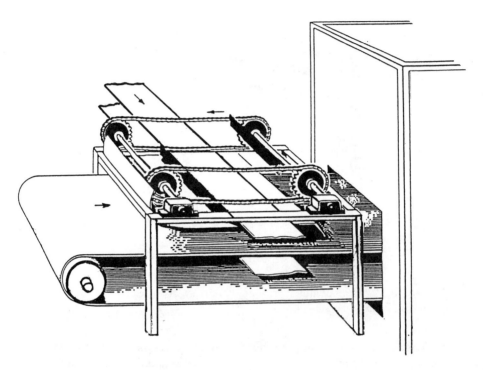

Figure 15.4 A wiper feed for a conveyor dryer (old version). (Courtesy of Proctor & Schwartz, Inc., Horsham, Pennsylvania.)

Figure 15.5 A rolling extruder.

Figure 15.6 A single-chamber pendulum lock (hydraulically controlled). (Courtesy of Salzgitter Maschinenbau GmbH, Münster, Federal Republic of Germany.)

15.7   Hazemag Rapid dryer

The majority of Rapid dryers are equipped with double pendulum self-sealing feed and dry product connection. The feeder introduces the material into the dryer and simultaneously it seals the dryer (see Fig. 15.6). How the feeder works can be seen by referring to Figure 9.8 in Chapter 9. Two pairs of flaps are mounted above each other, one pair opens while the other is closed. Thus, the material travels through the lock and enters the dryer. Normally, the pressure difference across the feeder should not exceed 100 mm wg (Motek, 1983).

It is interesting to note that flap gates cannot be used as extraction devices under bins or hoppers since the flaps would then have to close against the weight of the column of material resting on them. Two chamber gates are used to keep to a minimum the amount of air leaking into the dryer in unit time thereby reducing the explosion hazard. In this instance, hydraulic operation is better than mechanical. The feed and the dryer air pass through the dryer cocurrently and the feeder is thus, to some extent, exposed to temperatures. However, cooling by the feed stream also occurs. Pivoted flap gates are especially suitable for handling sticky, caking, or lumpy materials. The gates are made in sizes of up to 2,000 × 1,500 mm internal dimensions.

15.8   Convex dryer

The Convex dryer is shown in Figure 9.11 (see Chapter 9). Usually, feeds are free-flowing powders and are fed by means of a screw conveyor into the hot air channel just before the air enters the drying chamber. In some ways, Convex dryers resemble flash dryers.

It is also possible to process pastes and filter cakes which can be fed by Moyno pumps or twin-screw conveyors. In these instances, it is customary to install a mixer to disperse the feed. The dryer is now called *Pastes-Convex* (Blume and Matter, 1982; Buss, 1988). The dryer has been successfully applied to the calcination of catalysts, aluminum hydroxide, and zeolites.

15.9   Plate dryer

Plate dryers can operate under atmospheric or reduced pressure; in the latter case, special vacuum-tight rotary locks are used for the feed introduction and product extraction. Generally, rotary locks, kibblers, granulating screens, and table feeders are used for the introduction of the feed, although it is possible that other pieces of equipment may be preferred in some instances. Granulating screens perform, in a way, the same function as a roller extruder for conveyor dryers: it can form a cake into a type of extrudate that can be processed. *Case hardening* occurs when introducing relatively soft extrudates into the dryer. Kibblers disintegrate the feed to enable processing. Figure 15.7 shows a table feeder suitable for vacuum operation. The rotating table is present in the lower part of the feeder and is bottom-driven. The feed is distributed to the dryer through a peripheral hole. Baffles mounted on the table convey the feed to the chute.

15.10   Vigorously agitated contact dryer

Buss at Basel (Switzerland) manufactures vigorously agitated horizontal contact dryers equipped with an auger feed blade in the feed section. The system resembles their vertical thin-film dryer. Horizontal contact dryers can accept wet, free-flowing powders, and pastes that can be pumped. Figure 15.8 shows a feed system suitable for nonpumpable feeds that can be applied for continuous vacuum drying. The output of a batch centrifuge

Figure 15.7 A table feeder suitable for vacuum operation. (Courtesy of Krauss Maffei Verfahrenstechnik GmbH, Munich, Federal Republic of Germany.)

is received in a vibrating grid feeder and the feed enters the dryer via a pendulum lock. Feed shaping is not required since the dryer takes care of the formation of a thin film. Note: some filter cakes containing up to 60–70% water by wt can be made to flow under shear.

15.11 Vertical thin-film dryer and drum dryer

The feed for a vertical thin-film dryer has a low viscosity. It is customary to employ gear or Moyno pumps since these pumps achieve smooth dosing.

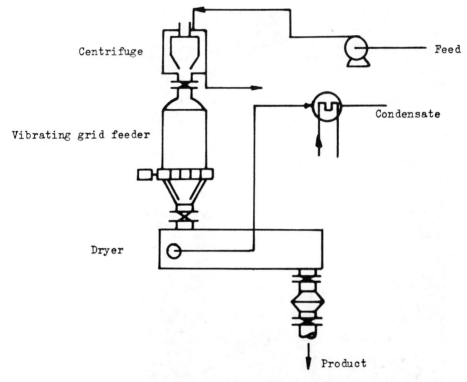

Centrifuge   Feed

Condensate

Vibrating grid feeder

Dryer

Product

Figure 15.8   A vibrating grid feeder. (Courtesy of Buss, Basel, Switzerland.)

The feed enters the top part of the dryer and is distributed by means of the agitator. The solution or slurry flows downward (Buss, 1988).

The feed to a drum dryer can be pumped into the nip between the two rolls, or by means of a top- or bottom-application roll. The feed can also be pumped into a trough and applied by dipping, splashing, or spraying.

References

Blume, G. and M. Matter (1982). CONVEX plants for the thermal treatment of inorganic raw materials, *Journal for Preparation and Processing,* 23, 219 (in German).

Buss AG (1988). Private communication.

Imes, P. V. and C. W. Jobes (1963). Fluidized drying and cooling of granulated salt, *Transactions of the Symposium on Salt,* Northern Ohio Geological Society Inc., Cleveland.

Komline-Sanderson Engineering Corporation (1988). Private communication.

Motek, H. (1983). Possibilities for using pivoted flap gates in bulk material preparation plants, *Journal for Preparation and Processing,* 24, 439.

Perry, R. H., D. W. Green, and J. O. Maloney (1984). *Perry's Chemical Engineers' Handbook,* McGraw-Hill, New York.

Proctor & Schwartz Ltd (1988). Private communication.

Vreeland, R. and J. H. Bacchetti (1982). Equipment plugging eliminated with fluid bed dryer, *Chem. Proc.,* 45, 24 (issue 12).

Index

Adiabatic saturation temperature, 65
Agglomeration, in suspension, 7
Agitated dryers, 198–203, 230 (*see also* Paddle dryers)
  calculation of, 201–203
  dust explosions in (*see* Dust explosions)
  pan type, 21, 34, 230
  process safety of (*see* Safety, of drying)
  under vacuum, 224–228
    calculation of, 226–228
    heat transfer in, 224, 227–228
    scale up of, 225
  with screw, 224–228

Backmixing, of product, 347
Baghouses, 328–333
  dust explosions in, 288 (*see also* Dust explosions)
Band dryers (*see* Conveyor dryers)
Batch dryers, 217–232
Belt conveyors, for feeding, 344
Belt dryers (*see* Conveyor dryers)
Biot's number, 131

Bulk density measurement, by radioactivity, 301

Calcining, 9
Centrifugal fluid-bed dryer, 262–264
Chute, 345–346
Conduction drying, 2, 35
Convective drying, 2, 35, 40–54
Convex dryer, 186–188
Conveyor dryers, 29, 174–183
  air-flow regime in, 175–177
  air velocity in, 177
  calculation of, 179–183
  conveyor velocity of, 177
  design method for, 178–179
  dust explosions in (*see* Dust explosions)
  feeding equipment for (*see* Feeding equipment)
  heat recovery from, 182–183
  layer depth in, 177
  multiple type, 175–176
  prices of, 178
  process safety of (*see* Safety, of drying)